Praise for *Leaving The Farm*

"This book is both a haunting elegy for a way of life that is fast disappearing and a beautifully crafted memoir about the universal experience of growing up. In an age of hype and hyperbole, Klatte remembers and reflects with self-effacing honesty. Using clear, elegant and evocative language, he has created an unforgettable work that delivers a powerful reading experience. *Leaving the Farm* is life-writing at its best."
—**Ken McGoogan**

"Ross Klatte sweeps the reader immediately into the excitement and fascination of childhood on a Minnesota farm. His loving attention to detail, and his consummate literary skill, takes the reader on a ride as wild as a toboggan run down a steep hillside alongside the barn. In fact, Klatte's masterful prose pulls the farm around us, so while the pages turn, the sights, smells, sounds, people, animals and landscapes of the farm become as much ours as his." —**Tom Wayman**

"Ross Klatte leads us to an epic comprehension of the loss of one family's farm, with writing so eloquent and disarming, so deftly nuanced and intensely moving that my sorrowful empathy with the tragedy herein is balanced by the sheer pleasure of reading such good writing. This is a wonderful achievement, one I'm sure Wallace Stegner himself would applaud." —**Caroline Woodward**

"*Leaving the Farm* brings to life a heartland agrarian lifestyle that has virtually disappeared in a single generation. Anyone who thinks that farm life is nothing more than the simple verities of crops, livestock and hard work, needs to read Ross Klatte. This is not a book of memories; it is a book of complex and luminous events." —**Don Gayton**

"A family history (which reads like a mystery), a journey of generations from Quebec to Minnesota to Saskatchewan and back, *Leaving the Farm* is the best of creative non-fiction, an intelligent driving narrative I couldn't put down."
—**Rita Moir**

LEAVING THE FARM

Leaving The Farm
Memories of Another Life

by
Ross Klatte

OOLICHAN BOOKS
LANTZVILLE, BRITISH COLUMBIA, CANADA
2007

Copyright © 2007 by Ross Klatte. ALL RIGHTS RESERVED. No part of this publication may be reproduced, stored in a retrieval system, or transmitted, in any form or by any means, without prior written permission of the publisher, except by a reviewer who may quote brief passages in a review to be printed in a newspaper or magazine or broadcast on radio or television; or, in the case of photocopying or other reprographic copying, a licence from ACCESS COPYRIGHT, 6 Adelaide Street East, Suite 900, Toronto, Ontario M5C 1H6.

Library and Archives Canada Cataloguing in Publication

Klatte, Ross
Leaving the farm : memories of another life / Ross Klatte.
ISBN 0-88982-237-9

1. Klatte, Ross. 2. Minnesota—Biography. 3. Farm life—Minnesota-History—20th century. 4. Farmers—Minnesota—Biography. I. Title.

F610.3.K43A3 2007   977.6053092   C2006-905252-2

We gratefully acknowledge the financial support of the Canada Council for the Arts, the British Columbia Arts Council through the BC Ministry of Tourism, Small Business and Culture, and the Government of Canada through the Book Publishing Industry Development Program, for our publishing activities.

Published by
Oolichan Books
P.O. Box 10, Lantzville
British Columbia, Canada
V0R 2H0

Printed in Canada
www.oolichan.com

*In memory of my sister Nancy,
lost on the farm;
and of the farm itself*

Author's Note

Except for close family members, the names of persons who appear in this work have been changed to protect their or their survivors' privacy.

Unless everything in a man's memory of childhood is misleading, there is a time somewhere between the ages of five and twelve which corresponds to the phase ethologists have isolated in the development of birds, when an impression lasting only a few seconds may be imprinted on the young bird for life. This is the way a bird emerging from the darkness of the egg knows itself, the mechanism of its relating to the world. Expose a just-hatched duckling to an alarm clock, or a wooden decoy on rollers, or a man, or any other object that moves and makes a noise, and it will react for life as if that object were its mother. Expose a child to a particular environment at his susceptible time and he will perceive in the shapes of that environment until he dies.
—Wallace Stegner *Wolf Willow*

"The Question Mark in the Circle" from *WOLF WILLOW* by Wallace Stegner. Originally appeared in American Heritage. Copyright © 1955, 1957, 1958, 1962 by Wallace Stegner. Copyright renewed © 1990 by Wallace Stegner. Reprinted By Permission of Brant & Hochman Literary Agents, Inc.

Contents

Overture

The Pepin Place
Imprinting / 25
The Pepin Place / 42
Pioneer Stock / 61
American Dreamers / 76
The Second Year / 89

The Mohrmann Place
The Mohrmann Place / 107
Sister School / 124
Wartime / 142
Postwar / 168
The Three Musketeers and Others / 185

The Woods and the Farm
Boyhood Eden / 199
Changes / 216
A Hick in High School / 233
The Tramp / 250
Summer / 262
Winter / 282

This 120-Acre Farm
Escapes / 301
When Things Were Going Too Good / 309
Class of 1953 / 323
Canoe Country / 332
The Accident / 348
Leaving the Farm / 358

OVERTURE

The day was clear and cold, one of those bright, biting winter days in Minnesota, below zero Fahrenheit with the sun shining. It was a day for tramping through the snowy woods or tobogganing down the modest hills or driving onto frozen Spurzem's Lake in my Model A Ford to put her into reckless spins past the critical, bundled-up farmers out fishing through the ice for crappies; it was a day, if you were my father, for hauling hay from our other barn in Hamel or letting the cows out to frisk in the yard or sacking up corn and oats to take to the feed mill in Hamel. But neither of us would do any of those things that day.

> *Having purchased another business,*
> *I will quit farming and will sell at Public Auction*
> *all my cattle, machinery, equipment and feed*
> *on my farm located 3 miles West of Hamel and*
> *3 miles East of Loretto—*
> *SAT., JANUARY 29TH [1955]*
>
> SALE TO BEGIN
> 11:00 A.M. SHARP

I have the old auction bill, yellowed and brittle now, and photocopies in case the original disintegrates. There, on that fading newsprint, is an inventory of the livestock and machinery my parents had accumulated after fifteen years of farming—years during which they grew from youth to middle age and suffered the loss of one of their children and I, their oldest child, grew from age five to twenty—years that were in another time, another place. It was another life.

54 HEAD HOLSTEIN CATTLE

We had got up as usual that morning, in the dark of six o'clock, and walked down to the barn, our rubber overshoes squeaking on the dry snow, the cold stars still bright overhead. Dad went on into the cow barn, closing and hooking both halves of the Dutch door from the inside, while I turned into the milk house, the small brick building attached to the barn, to prepare the milking utensils.

> *3 year old heifer, fresh Oct. 15, bred back Dec. 15*
> *2 ½ yr. old heifer, fresh Nov. 15, bred back Jan. 15*
> *Holstein cow, calf at side, cow registered*
> *Holstein cow, fresh Dec. 1st, open*
> *Holstein cow, fresh Jan. 15, calf at side*

I drew steaming water from the electric water heater, added disinfectant, and rinsed the separated parts of our three Surge milkers. Then I assembled them, metal buckets and covers, metal-encased rubber teat cups, finally the pulsators that, when connected to the vacuum line in the barn, would cause the cups to squeeze and pull at the cows' teats, mechanically substituting for human hands.

There were several ten-gallon milk cans to rinse, two iron milk pails, the strainer, which I then fitted with a couple of "pads," cotton filters that when dirty would be replaced during the milking.

> *1954 Allis Chalmers WD 45 tractor, fully equipped with fluid wheel weights, narrow front end, quick hitch, 3 bottom 14-inch plow and 2-row powerlift cultivator*
> *International F-20 tractor on rubber in good condition*
> *1949 New Holland baler*
> *1952 Allis Chalmers 6-ft combine with Scour Clean*
> *1952 Wood Bros. corn picker*
> *1949 Massey Harris 7-ft. power mower*

How many winter mornings have I worked despondently here, depressed by the cold and the clouds of my own breath, longing to be somewhere else—how many pre-dawns and dark-at-four-o'clock afternoons of my teenage years have I given, without hope of escape, to these dreary chores? Yet that morning, buoyant with relief and excitement, heavy with guilt and premature homesickness, I perform each action, each step in the routine, as if I'm doing it for the first time. That's because it's the last time, I keep reminding myself. After today, I am leaving the farm.

> *Holstein cow, due to freshen Feb. 5*
> *Holstein cow, registered, fresh Jan. 2, bull calf at side*
> *Holstein cow, fresh Dec. 15, bull calf at side, Reg.*
> *Holstein heifer due to freshen Feb. 10, 3rd calf*
> *Holstein cow, to freshen May 1st, registered*

All those purebred Holsteins, "fresh" or due to freshen (that is, calved or about to, producing milk or about to—or "open," not yet bred back); all that late-model machinery. All listed, painstakingly, perhaps lovingly, by my father at the kitchen table one night, his stubby fingers around a stubby pencil, to be set up in print so that it reads like blank verse on the auction bill. It's a summing up of the folks' investment in a way of life that was my father's stubborn hope and my mother's nightmare, finally. While between the lines I read how the expense of mechanization and the vagaries of the economy, my leaving, and a death in the family brought an end to that life.

Along with the auction bill I have a bundle of old photographs, rescued from a dusty shoe box found in the garage of the folks' house outside Minneapolis and brought here to British Columbia where, in 1970, my city-bred wife and I moved "back to the land."

The box was labelled "Pictures of the Old Farm," and nobody had much looked at them until, after my father died in 1987, my sisters and I, with our mother, rummaged through the box. I doubt my father ever looked at the old pictures, though he used to reminisce, compulsively, not only about the farm but of his struggle through the Great Depression when he ran a filling station, and how he went into farming just before the war, and how eventually he started a golf course on what was left of our old farm. *He.* My mother, in his telling, was pretty much in the background, though he counted on her support; they supported each other.

Your father had such a temper, but he was a good provider. He was a good man.

Instantly, the old snapshots take me back to the farm, views suddenly so familiar again I want to step into them, return for a moment to that other life, that time and place of so long ago. Nostalgia. Yes, but something more, a recognition of my early self and where I came from, a sense of that lost place of my origin which I can't, don't want to, go back to anymore, just as I can't and don't want to go back into the Navy: I've served my time.

There are shots of the old barn, big and painted yellow like all the buildings, though their color doesn't show in the black-and-white pictures; the two-storey chicken coop that burned down one night, ending the folks' egg business; our old farmhouse, uninsulated except for the attic the last year or so, drafty in the winter, its jigsaw trim under the eaves broken where I used to swing from it as a child.

And pictures of the folks when they were young: my father dark and skinny from working in the fields all summer, my mother pale, fat from house-bondage and "the blues" and being pregnant too much of the time.

Pictures of my sisters and brothers and me as small children—as kids in grade school—and finally, Joyce and I, as self-conscious high school teenagers.

The land. The land just emerged, bare and dismal, from the snow; then worked up for planting; then covered with oats or alfalfa or rows of corn.

Our cows in pasture. A family picnic in the back yard.

Nancy.

❧

I rap on the barn door. Presently Dad unhooks it from inside and I step into the moist rush of animal warmth.

Arms loaded and in a couple of trips, I carry in the milking machines, the milk pails, a pail of warm, disinfected water with a rag in it for washing the cows' teats and a pail of cold water for rinsing the teat cups. I attach the vacuum line hoses to the pulsators and the milkers are ready.

Dad steps down from the manger, where he's been rationing ground feed to the cows. You see his limp now: the result of losing most of his left foot in a silo filler in 1943. It was cut off, to the instep, and he stuffs old socks into that shoe to jam the stub against. He can feel his missing toes, he says; they ache most of the time, especially before a weather change, before a rain.

"You can give'm silage," he tells me.

He starts the milking. I feed the cows silage, hauling them big forkfuls of the mushy, fermented stuff, chopped corn I threw down from the silo the night before. They plunge into it like drunks into schooners of beer.

Then I go out again, latching the Dutch door from the outside, and walk uphill and around to the loft to throw hay down. I lift the bales out of the mow, the good second or third-crop alfalfa (my father's pride: none has been rained on, thanks to his aching foot), and drop them down the chute to the manger. Then I walk around to the cow barn again, break the bales and spread them.

> *Van Brunt grain drill, 20 disc, with grass seeder attachment*
> *Rubber tired wagon and rack*
> *David Bradley corn planter with fertilizer attachment and check wire*
> *International manure spreader on rubber*

I open the door to the barnyard and pull the manure carrier in on its overhead track, lower it, and start cleaning out the gutters. They're filled with the overnight mess of shit and piss and straw binding that I lift out with a scoop shovel and plop into the carrier. How many times have I cleaned the barn over the years, twice each day in the winter, morning and evening, and how many carrier loads (three of them each cleaning) have I pushed outside and dumped onto the manure pile, the pile growing until, by spring, it was as big as a building. In the spring, as soon as the snow was gone, you had to haul the manure out to the fields and spread it, load after load, until the pile was gone, before the milk inspector came.

Crouched beside a cow with a milker on her, my father pinches the hose under each teat cup to test for milk still flowing into the bucket. When the flow has stopped, he stands, turns the vacuum line nozzle off, pulls the hose attached to the pulsator off the nozzle and loops it in his hand. He eases the teat cups from the cow, lifts the support belt off her and slings it over his shoulder, then steps onto the walkway lugging the bucket. He lifts the cover and pours the warm, frothy milk into a pail that I will dump into one of the ten-gallon cans in the milkhouse. Then he dips the cups in the rinse water and moves to the next cow in line.

"Hurry up with the barn," he says. "You got calves to feed."

But there isn't the old, angry tone in my father's voice. His anger is spent. Things are out of his hands now. There's been no real anger, no belittlement of me, no exasperated pleas to *Take an interest, for chrissake*, since the accident, really, since he knew his oldest boy was going to leave, since his wife was ready to quit, since his own will to keep farming had seeped out of him.

"Think they'll take all the cows today, Dad?"

"Naw. By Sunday night, though, they should all be gone."

Sunday night I'll be at an aunt and uncle's house in Minneapolis, sleeping fitfully on their living room couch. The next day I'll be on a train, headed for Navy boot camp.

But the farm will go with me, its grounded reality, the daydreams it inspired. I won't forget its drudgeries and, yes, its satisfactions; the tension between its riches and deprivations. I'll be freighted with memory—memories of my former life on the farm and of my parents and sisters and brothers and my Catholic upbringing and how I balked at what my father had to teach me. His verbal abuse. Family jokes, and family sorrow. My exposure to nature, to all the wild places around the farm, the woods and marshes and little pothole lakes that gave me an escape, became part of an imaginary world I created and inhabited so fiercely it took an act of will finally, a kind of amputation, to throw it off. *Wake up*, my father used to holler. *Pay attention.*

> *Miscellaneous small tools too numerous to mention*
> *1942 Dodge truck with very good grain box*
> *1929 Model A Ford car*
> *Piano*

> THIS 120 ACRE FARM FOR SALE BY OWNER

1

The Pepin Place

IMPRINTING

Whenever I hear the young Judy Garland singing "Over the Rainbow," from the 1939 movie *The Wizard of Oz*, I get a sinking feeling, an unease out of the shadowed recesses of memory that in part is a response to the song's yearning and the lovely, heart-felt quality of Garland's voice—and in part is something else. That something else has to do with a tornado, which I saw or felt the presence of when I was four years old, linked with the movie and its simulated tornado, which I saw sometime later.

I can date the tornado exactly. Among my old pictures are two faded snapshots taken on the very day of the tornado. One is of my sister Joyce and me, Joyce on a tricycle, me standing beside her, taken on our grandfolks' lawn in Hamel, Minnesota. The other picture is of a nearly flattened V-8 Ford sedan, dropped into the middle of a farmer's field. At the top of the picture is a one-word explanation in my mother's hand: *Tornado*. Under both pictures is scrawled the date: *June 18, 1939.*

That was a Sunday, a Sunday after church and then dinner at the grandfolks' farm in Hamel. The day had been hot and still, breathless, with that stupendous gathering of clouds which warns of violent weather. *The sky was green*, my mother remembered.

Suddenly we kids were shooed into the cellar. The men climbed to the second-storey landing of the house and looked northeast over the Soo Line tracks below the town.

"You see that, Bert?" I heard my father say excitedly to my grandfather. "Look at that!"

"*What?*" my mother called up. "What do you see?"

I poked my head out of the cellar, craning to see for myself, but there were sheds and trees in my way. The sky above was serene.

If I never saw the tornado, did I hear its roar as it dropped out of the whirling clouds overhead and touched down, raising dust, shredding trees and buildings, beyond the town? It churned through the countryside toward the Mississippi River and the town of Anoka, where people were killed, we heard.

After the storm all of us, kids and adults, set out to see the damage. I rode in somebody's black, hearse-like car that took us down the depot hill, across the tracks and north up 101. We passed the ruins of a barn, with only its stone walk-in basement, the cow barn, left standing. We saw the car out in a field, crushed into scrap metal, and somebody took a picture of it. We stopped and looked down at the body of a woman, lying beside the road with a leg missing. She had been in the car, somebody said. The air was fresh and cool and birds were singing and I felt sick.

Some where
o-ver
the rainbow

Judy Garland sings, and that old empty feeling, a sense of all the beauty and terror of life, comes back. It's like the distant wail of a train passing in the night, at once lonesome and alluring. And maybe it's linked, somehow, to my mixed feelings about where I grew up, the place I left and that has never left me.

Late spring or early summer of 1940. I'm five years old. The folks pile my younger sisters and me into the back of our family car and we all go for a drive in the country.

My father drives us out of Hamel, the village some twenty miles northwest of Minneapolis where my mother grew up and that was named after her grandfather, the son of nineteenth-century French Canadian immigrants from Quebec. We lived in Hamel then, still mostly populated by descendants of the original Quebecois immigrants, on her parents' failing farm on the edge of town, after moving there the year before from Minneapolis, where our father ran a filling station and my two sisters and I were born.

Seated in the folks' 1938 Chevrolet with its stylized, diagonal crease down its sides, we move along the town's main, unnamed street, past St. Anne's Catholic Church, past the DesLaurier farm at the west end of town, and on down the hill past LaMere's Garage.

Driving over the potholed blacktop of the Rockford Road—soon to be called the Old Rockford Road, since Highway 55, dedicated as Olson Memorial Highway for Floyd B. Olson, the former governor of the state, will replace it that year—we pass farms and fields, swamps and pastures, stands of hardwoods. The crops are up, alfalfa starting to blossom, the wheat and oats, not yet headed out, like rows of grass, and the rich Minnesota soil black between the cultivated rows of young corn. With the car windows open, we hear the tinkle of meadowlarks, the reedy call of blackbirds, see cows and horses along the fences.

Then we drive between thick woods on our left and wooded pasture on our right, the trees seeming to arch over the road, and pass a place on our left where two lanes begin, one going straight into the woods and the other, with a mailbox at the head of it, disappearing obliquely into the trees. We break into the open and my father stops the car, turns it around, and pulls off the road on the edge of a bank that overlooks a willow swamp. Beyond it, there's rolling farmland.

"There it is, kids," he says, pointing out of the car windows. "Our new home."

"Where? *Where?*" Joyce and I cry, staring out from the back seat of the car. Marcia, hardly a year old then and sitting up front on my mother's lap, must have gawked out the window too.

"There."

He points below the road and across the swamp to a little farm. It's like a calendar picture of a farm: red barn, metal shed, a brown house surrounded by shade trees. And a windmill, its galvanized metal derrick poking up above the cluster of trees and buildings.

"See it? That's where we're going to live."

"When? *When?*"

"Pretty soon."

❧

A week or two later we drove out from Hamel again, turned into the driveway through the woods and came out where there was a big field on our left and a meadow pasture on our right that sank into the willow swamp you could see from the road. The driveway ran into a dirt yard encircled by the farm buildings.

The folks, with money from the sale of their house in south Minneapolis and my father's filling station in north Minneapolis, had bought this place—house and outbuildings and forty acres of land, including pasture for the thirteen cows they would have to truck over from the grandfolks' farm. The previous owners, the Pepins, had bought a tavern in Hamel, the one, in fact, across the street from the grandfolks' house that Grandpa, despite Grandma's disapproval, liked to sneak over to for a beer and a shot of whiskey between farm chores.

The small house was sided with brown shingles and screened from the yard by tall lilac bushes. The high front porch above the walk-in basement overlooked a patch of shaded lawn and a barbed wire fence between the smooth grass of the lawn and the grassy humps and depressions in the meadow pasture. Those bumps, I knew, came from a cow's hooves sinking into the pasture's soft ground, and where the grass grew longer and greener in circular little bunches, that was where the cows had fertilized it with their "pies."

The barn was smaller than my grandfather's, but its roof extended to within a foot or so of the little milk house that stood between the barn and the windmill. Eventually I would brave the ladder on the windmill to reach the roof of the milk house and from there jump over to the barn roof. I would climb the windmill, too, someday.

Inside the barn it was dim and cool. There was a double row of empty stanchions with mangers in front and a cement walkway between the gutters. At the end of the walkway, through the open top half of the Dutch door, you could look out into the barnyard and beyond to a rise of land and the curve of a hillside pasture above the willow swamp. Left of that, along a line of trees, was a field road leading up over the rise to the back end of the farm. On the other side of those trees was the big field that stretched along our driveway and the south border of our farm. It belonged to another farm, whose buildings, though less than a mile away, were out of sight below the level of the field. That larger farm was where we would move to in a couple of years.

"The Brown House," Joyce and I called this place. To our folks it was "the Pepin Place." It was here that I was first exposed, as Wallace Stegner has said of his "susceptible" years in the Cypress Hills region of Saskatchewan, to that environment in whose shapes I will perceive until I die.

Mornings in the Hamel house I'd climb the stairs to the second floor where, having surrendered the first floor to us Klattes, the grandfolks lived. I'd find Grandma at her breakfast.

"*Bon jour!*" she'd say. "*Comment ça va?*"

"*Tres bien.*"

It was about all the French I would learn from her. That and a few other words such as the expression for "little fart," her term of endearment for me. And she taught me how to say Hamel in French, *ah-mel*, the silent "h" and the stressed second syllable making it another name entirely.

French was my grandfolks' private language by then: they would break into it at times, excluding us kids and most of the adults around them. That wasn't the case when my mother was a child. She didn't speak English, she told me, until she went to the one-room public school in Hamel; later she continued to say her prayers in French but eventually "forgot" her ancestral language, though she understood it, I think, for years after losing her ability to speak it.

It was always summer in those mornings upstairs in the Hamel house with the delicious smell of toast and coffee and the sun warm on the kitchen table by the east windows; Cedric Adams on the radio and Grandma's immense, commanding figure. She ruled the house. It was a house full of women, of girls, including my mother and my sister Joyce, my aunts Clarice and Bonnie, one a teenager, the other only five years older than I was. Grandpa Hamel and my father, out on the farm or at their other jobs, were hardly ever, it seemed, in the house. My grandfather may have been working for the highway by then. My father worked nights at the Coca-Cola bottling plant in Minneapolis.

"Sit, little fart," Grandma told me in French.

So I sat at the table and watched her make toast and hot cocoa for me (the cocoa served in a heavy mug with a picture of Shirley Temple on it) and listened to her talk. She spoke in quick, Gallic bursts, showering me with warmth like that from the sun streaming through her windows.

She used to pinch my cheek or suddenly sweep me up and bite the lobe of one of my ears until it hurt. But I knew she loved me, knew it even when, gently but firmly, she broke me of my childish need to be completely naked to do "number two." Perhaps I'd been toilet trained too early; my mother bragged that I was "completely clean" at a year old. I had a fear, anyhow, of soiling my clothes and so had to have everything off, including my shoes and socks, before I sat on the pot.

C'mon. Do your job now.
I can't, *Grandma. I* can't!
Sure you can.

Standing over me, hands on her hips, as I sat miserably on the pot in her upstairs hallway, fully clothed with only my pants down. Straining. Finally doing it.

I was rewarded with a stick of gum: Wrigley's spearmint.

At night, outside the window of the downstairs room where Joyce and I slept, there'd be the sounds of the town, loud voices from the tavern across the street, drowned suddenly by a car starting up to rumble away out of hearing. And then the wail of an approaching train, and the clicking, metallic roar as it tore through town like a passing storm. And then the stillness afterwards, just the chirping of crickets under our window or the sighing of the wind through the trees around the house.

The grandfolks' big clapboard house (built in 1907, at the height of the Hamel family's prosperity) had lilac bushes along the driveway and Grandma's flowers off the columned veranda. In back of the house, in a kind of hollow sloping to the edge of the bluff above the Soo Line tracks, was her vegetable garden with a couple of apple trees at the head of it. There was a pump house out back and a line of sheds leading to the barn. The front lawn ran down to the dusty path between the cement sidewalk going uptown and the filling station at the head of their driveway where the attendant, with little business to attend to, used to sit outside and read the paper.

Albert Hamel, my grandfather, liked to drink and play cards. He smoked roll-your-own cigarettes too, had a big French nose peppered with blackheads, and though sometimes crabby, more often he had a joke to tell. There was a song he liked, "Cigarettes, Whiskey, and Wild, Wild Women." Just mouthing those words brought a chuckle from him. Not that he indulged in the "wild women" part, though he had, according to Grandma, been "wild" as a young man, courting girls in his family's rubber-tired buggy—better yet, in the family cutter on cold winter nights when you could snuggle together under the blankets. In a sleigh or buggy, you could neck with the girl all the way home. *The horse knew the way.* It knew, even, to pull over when meeting another buggy or sleigh to avoid colliding with it.

You young fellas in your cars nowadays. You don't have nothing over what we had with a horse and buggy!

Grandma was the former Emily Mingo, born in the old French-Canadian community of Little Canada, near St. Paul, and raised northwest of there in Elk River, where, one winter evening, she met Albert Hamel at a dance. He asked to take her home. Her older sister Ellen, also at the dance, knew about Bert Hamel and his way of loosening up a girl by running his horse too fast around a bend so as to tip the sleigh over in the snow. *Ellen warned me: Don't go out with Bert Hamel, he'll tip the sleigh!*

She went with him anyway. And, sure enough, he tipped the sleigh.

They were married in 1910. She was a couple of years older than he was.

❧

Hamel then was a village of three hundred inhabitants. There was the one paved street, which was the Old Rockford Road out of Minneapolis as it passed through the town, and a few dirt side streets where the fields met the backs of the houses. Uptown was the Catholic church, a great high, yellow-brick structure with a square Italianate tower that only six years before had replaced the original wooden church built in the 1860s. From it a bell tolled at eight each morning, at noon, and at six o'clock, marking the hours.

Four taverns. Two grocery stores. A garage and a filling station. An old hotel that I watched burn down the summer of 1939, and an old, two-storey hall from which oompa music still carried over the town on weekends from its upstairs dance floor but whose glory days had been earlier in the century when relatives used to come out on the train from Minneapolis for the weekend dances. Afterwards, they slept in the spare bedrooms or on the floors in my grandfolks' big house.

A one-room school, and next to it the tiny building housing the Hamel branch of the Hennepin County Library. Feed mill, hardware store, blacksmith shop. The Farmers' State Bank of Hamel.

There were boys my age in the town to play with. We used to get naked together, behind my grandpa's shed. Once during a peeing contest, after squeezing my penis and holding back for maximum pressure, I let go and was abruptly blinded and completely disoriented. I'd pissed up my nose!

After the joy of being naked, though, came the sorrow. I couldn't dress myself yet, and had to go crying, clothes in hand, to Grandma for her scolding help. I liked to sleep naked, too, which my grandmother quickly discovered. At night she would stick her head in the room where Joyce and I slept and cast a knowing eye at me. "Are you *naked?*" "Yaas," I'd have to say. That would cause her to dress me in my pajamas again, but after she was gone I'd again take them off only to wake in the morning with my pajamas on once more. How had that happened?

Diagonally across the street from the grandfolks', next to Pepin's tavern, was the combination beer joint and butcher shop owned by Johnny Cavanagh. He was a town character, an unabashed opportunist out of the raw old days of American free enterprise and about whom there were funny stories. Like the time a man came into his store to announce that so-and-so's cows had got on the Soo Line tracks and a train had plowed through the animals, killing a couple of them. "Where'd that happen?" Johnny asked, grabbing his meat saw and butcher knife. By the time the owner of the cows arrived on the scene, Johnny had the dead animals skinned, cut up, and loaded onto his truck. "I was here first!" he told the dumbfounded farmer.

His grandson Ducky and I used to play together behind the Cavanagh place, in and around Johnny's barn where he kept his meat animals. One day, utterly fascinated—and completely ignored by him—we watched the old man butcher calves.

The calves were led, one by one, out of the barn and tied to a tree. Then struck down with a hammer blow to the forehead. Stunned and quivering, the calf's head was then pulled back and its throat cut. Sprays, torrents of blood. Hind legs kicking. Then the body was strung up with a block and tackle and gutted, the hot (really hot! Ducky and I touched them) entrails spilling out in a pile. Then the skinning, the raw carcass emerging from the hide. It was all done without squeamishness or hesitation, in quick, grunting displays of the man's skill.

There was a sign of a flying red horse over my father's filling station before our move out to Hamel. I have an image of him bustling around a customer's car, washing the windshield, working the pump to fill 'er up. The gas business was a good one to be in during the Depression.

Christ, people had cars and kept driving them no matter what. They'd buy gas when they could hardly afford to buy groceries!

At the time of his marriage to my mother, in November 1933, he was a pump jockey for Standard Oil. Then, some time after the Social Security Act of 1935, when the oil companies, to pass on support of the plan to their employees, began leasing their stations, my father got his own Mobil station on the corner of Broadway and Lyndale, in north Minneapolis. With a helper, he stayed open around the clock; and started, in the teeth of the Depression, to make money.

I sold tires, I was in the used tire business. I sold live bait. Gave the daily paper to customers with a fill-up of gas. I sold pop—my station was the first in Minneapolis to have a cooler. Anything to make a buck, to keep'm coming back. Service. Standard Oil taught me that.

Twelve-hour shifts, though, seven days a week. We had so much business on Saturday night, I'd work straight through with my helper. That's how I got ahead.

His first month, Mobil's main office in Kansas City wrote to the Minneapolis office: "You must have combined two months in your report for Station 42."

No, Minneapolis wrote back, it was just that my father was now the operator.

Mine got to be the best Mobil station in Minneapolis!

When the house in which my father grew up was going for taxes, my folks, who'd been renting the upstairs of my aunt Ellen's house in north Minneapolis, paid the taxes, got the house, and allowed my great-grandfather Williams, who had built the house, to stay on after we moved in. It wouldn't be the last time my folks helped a relative, and themselves, during the Depression.

That house, at 1933 Drew Avenue (a number easy to remember as 1933 was the year my parents were married), caught fire one day, which brought a wailing fire engine and firemen in black helmets and raincoats who clambered over the roof of the house to chop holes in it. Why were they doing that, I wondered? It seemed they were helping to destroy the house.

After the fire, my father, with his filling station helper, rebuilt the house. Actually, it's the rebuilt house I see most clearly, its striped awnings and stuccoed sides, the roof angling down to the arched entrance into the back yard, mental images only confirmed by a snapshot, dated July 28, 1938, of me, aged three and a half, and my older cousin Barbara, standing barefoot in sunsuits on the front lawn.

I had a little pedal car that was stolen one night when I left it out on the sidewalk at the bottom of the hill, too tired to push it up. I climbed the hill to the house and asked my mother for help, but she was at her desk—writing cheques to pay bills, probably—and sent me away. The next day the car was gone, of course, and I was told, "See? That's what happens when you leave things out." I felt wronged but swallowed it.

There was a grocery store at the bottom of the hill where the owner gave me penny candy, and the lake, Cedar Lake, which I could see gleaming at the end of the tree-lined street and which I never went near because I might drown in it, my mother said. When the ice truck came, the man chipped slivers off a block of ice for my friend Snooky, who lived next door, and me to suck like popsicles before carrying the block with tongs into my mother's kitchen and lifting it into the icebox.

One day I stood on the sidewalk below our house and watched my father, once a golfer who no longer had time for it, hit a ball over the empty lot across the street. It whizzed over my head and disappeared into the sky.

Then: *The war started in Europe. I knew it was only a matter of time before we got in it.*

Meanwhile, there was a "gas war" between the filling stations in the Twin Cities, and certainly, when war came, there'd be gas rationing. Besides, there was a draft already. It seemed to my father a good time to go farming. Hadn't he married a farm girl? Anyway, more pragmatist than patriot and approaching thirty with a wife and three kids, he doubted he'd be drafted—especially as a farmer.

I figured no matter what happened, we'd always eat, and for sure the country'd need milk. Besides, we might even make some money.

Never mind that going into small farming at that time was against the national trend. By 1939 the days when a man could settle on a piece of land, farm it, and make more than a subsistence living were about done, though few, least of all my father, would admit it.

Starting in the nineteenth century with the invention of the mechanical reaper and the sod-busting plow, and accelerating through the teens and twenties and into the thirties of the twentieth century when, along with the worst economic depression the country had known—bank failures, dust storms, mass foreclosures, mass migrations to the golden land of California—mechanization increased, and the so-called family farm began giving way to big business. World War II would slow the process—provide artificial respiration, give a shot of temporary prosperity to the small farmer—but mechanization would pick up again after the war and eventually turn farming into agribusiness.

But who knew this then? Certainly not my folks, who sold our Cedar Lake house in the spring of 1939, and moved us, including my great-grandpa Williams, pushing ninety by then, in with the grandfolks in Hamel. The old man was given a cot in the upstairs hallway of the Hamel house.

Your grandpa's barn had stanchions for twenty-six cows. He was milking only thirteen at the time, and he said, "You buy another thirteen cows and we'll fill the stanchions. We'll milk'm together and split the milk checks." And I thought, hell, that's a way to get started. But I had my night job at Coca-Cola and so I never took my half of the milk checks. I let the grandfolks have it. That saved their place! They hadn't paid their taxes in seven years!

I watched the milking. I liked the dimness and warmth of the grandfolks' low-ceilinged, animal-heated barn in the evenings. My father and Grandpa Hamel sat on little stools between the cows and squeezed milk from the cows' teats into metal pails, the sound high and metallic at first, then low and muffled as the pails filled with warm, foamy milk under the kerosene lamps hung from a wire stretched along the ceiling behind the cows; you moved your lamp as you moved down the line during the milking.

"Open your mouth!" Grandpa would say. And he'd squirt warm milk at me, a white explosion in my face. He'd squirt at the several cats in the barn, too, causing some to sit back and simply take it and others to leap into the air and paw at it as if trying to catch a bird in flight.

THE PEPIN PLACE

Our first night on the Pepin place, Joyce and I were put to bed upstairs, in the "unfinished" part of the house. It was really the attic, with bare-studded walls, exposed rafters under the peaked ceiling, and cobwebs in the upper corners. The smell of dust. There were small windows at either end of the long, narrow room, and an old bed, the only furnishing, in the middle of the floor.

No electricity—not many farms had it then, but the folks would get it soon under the REA (Rural Electrification Administration), part of Roosevelt's New Deal. So Joyce and I were escorted upstairs with an oil lamp that cast huge shadows on the raw walls and accentuated the spookiness of the place. There must be bats up here, I thought. They got in your hair, according to Grandma Hamel.

"Bats!" my mother said. "There's no bats up here, are there, Mervin?"

"Naw," our father said. "You kids'll be all right."

Just in case, I clapped my baseball cap over my head. My mother smiled at that.

The stairs creaked as the folks left us. Our father called up, "Good night, kids, sleep tight!" and our mother added, "Don't let the bedbugs bite!" Then they closed the door at the bottom of the stairs and Joyce and I were left alone in the creepy dark.

I was still awake, staring up at the darkness, when I heard the car start up outside and my father drive away to his night shift at the Coca-Cola bottling plant in Minneapolis. Beside me Joyce was asleep, her long dark hair spread trustfully across her pillow.

I must have slept too for a time, then woke suddenly to stare up into the darkness again. Something had awakened me. Then I heard it: tiny, high-pitched squeaks and the faintest scurrying, directly above me among the rafters.

I pulled my cap down over my forehead and the covers up to my chin and tried to see in the dark. My naked face, my eyeballs even, felt dangerously exposed.

Then something, like the faintest breath, swept past me. Simultaneously I realized I could see the far window, dimly illuminated by the outside light of the stars or a partial moon. Staring at the window, I saw a bat fly past it. A *bat*!

I dove under the covers.

"Joyce!"

I nudged her, but she only kicked at me, snuggling deeper into sleep.

"*Joyce!*" I whispered frantically up through the covers. "*Wake up!*"

She stayed asleep. I came up for air, saw more bats, dove under again and, feeling protective of my younger sister, tried to pull Joyce down with me, tried to cover her exposed hair. But she pushed me away.

I imagined a bat landing on the bed covers, felt its tiny weight, its clinging little claws, scuttling along the top of the blanket. I kicked at the blanket and then was certain that the bat had slipped under it. *It's under the blanket with me!*

That's when I abandoned my little sister, slid out of the bed, and flattened myself against the floor. I began to crawl, very slowly, toward the stairs. I wanted to leap up and run! But I willed myself to crawl. At the stairs I slithered down them until my head was below the floor, then stood up and felt my way down the rest of the stairs to the door and opened it. I found the folks' bedroom and woke my mother.

"*Mama.* Mama! Wake up!"

"Huh? What's going on?" The bed creaked and I saw the dark outline of her raised head.

"*Bats*, mama. *They're upstairs!*"

"Shhh. You'll wake the baby." Then, accusingly, "Did you leave your sister up there?"

She went upstairs with a flashlight. I heard her creaking across the floor up there, apparently ignoring the bats. Downstairs again, she came into the bedroom carrying my sister and placed her next to me. Then crawled into the bed with us. Lying next to her warm, reassuring bulk I saw again, in my mind's eye, those flitting bats; but now I could watch them from the comfort and safety of my mother's bed. Yet something, some vague unhappiness, began to nag at me. I was trying to think what it was when suddenly it was bright morning and my father stood over me, home from the Coca-Cola plant, in his farm clothes already.

"You awake, Ross? Wanna help me?"

"Yuh!"

"Get dressed then. We'll get the cows."

Out in the pasture a cow rushed at me, blowing through her nostrils, and my father called from across the herd, "Yell! Wave your arms!" So I yelled in my high-pitched child's voice and waved my arms, and the cow turned away. She started plodding toward the barn and the rest of the herd fell in behind her. My father stood watching them as I ran up beside him.

"You see how when you get the leader going the rest follow?" Studying me, my father said, "You get a little older, I'll send you out alone to get the cows. Think you could do that?

"Yuh!"

"Good boy."

In the yard my father moved through the bunched animals and opened the door into the barn. The cows jostled one another, competing to get inside. My father looked across their backs at me, grinned, and I knew that more than anything else in this world, I wanted his approval.

We got Oscar. He'd been my grandfolks' hired man in Hamel, became like their adopted son, then after the grandfolks couldn't afford him anymore he found work outside Wichita. There he met a woman *willing to marry me*, as he wrote "home" to the grandfolks. *Don't do it*, was Grandma's advice. *She's only after your money.* What money could Oscar have earned as a Kansas farm hand? Anyway, he turned up at the grandfolks' again, and they sent him to us.

He was still a young man, in his early forties maybe, but he smoked too much, and he'd had rheumatic fever as a child. He wasn't as strong as he should have been.

Yet he was hard-working, though a little simpleminded; a gentle, lost soul, my mother recalled, who needed taking care of. *He'd had a hard life. I don't think he was treated right as a kid.*

Good old Oscar, people said of him. He had a rich, intoning voice, which I was reminded of the first time I heard a recording of Carl Sandburg reading his poetry. It was like the sound of a harmonica. Did Oscar play what we called a mouth organ?

He smelled pleasantly of sweat and tobacco, alcohol sometimes, and didn't wash much more than his hands and face until my mother made him take a bath. My father's smells weren't as strong; he didn't drink or smoke and, like us kids, took a washtub bath every Saturday night. But like Oscar, he sometimes resisted, in which case my mother laid down an ultimatum. *I'm not sleeping with you unless you take a bath.* That caused Oscar and my father to grin at each other. There were things you had to do for a woman.

Having Oscar made it easier for my father to keep working nights at Coca-Cola. He'd get home in time for the morning milking, then have breakfast and work till noon with Oscar. Then he'd sleep through the afternoon while Oscar carried on with the farm work. He'd be up for supper and evening chores.

After chores, my father might go to bed for another hour or so, then be up again about the time we kids were put to bed. "So long, Ruthie," he'd tell my mother. I'd hear the car pulling out of the yard. I'd see him come back in the morning.

Joyce and I slept downstairs now, on the opened studio couch in the parlor of that small house. Marcia slept in her crib in the folks' bedroom. Oscar slept upstairs with the bats.

❧

A warm bright summer morning. Joyce and I are playing in the dirt driveway. My mother calls from the house.

"*Let's Pretend* is on!"

Let's Pretenders, to be exact, came on the radio every Saturday morning at ten o'clock. It dramatized fairy tales like "Rumpelstiltskin" and "Hansel and Gretel," two of my favorites because they spoke of "the forest"—the great forest of medieval Germany, I would learn, out of which came these folk tales. We had a forest of our own, the woods through which our driveway ran and that stood at the east end of the big field past our farm buildings. I imagined these woods inhabited by Old World witches and goblins and was afraid of them at first, especially since my mother warned: *Don't go out in the woods by yourself. You'll get lost!* One day I went into the woods with my father and Oscar, and discovered their other world.

It was an early summer evening—most likely after chores—and I followed the men across the field beyond the farm buildings to the woods. The field was planted in corn—the corn just up, little shoots of green in the dark soil—and was like a dark pond, with the woods growing thickly down to the water.

My father carried his shotgun. He meant to shoot the crows that raided the field every day, fed on the shoots of corn, and roosted in the woods every night. I saw and heard them from the farm in the late evenings, the long black undulating line of them, as they flapped, cawing, over the field toward the woods.

We reached the woods, stepped into the trees, and it was like entering a huge empty building, silent, secret, alive. The trees stood like columns, with park-like regularity, the trunks going up and up to the spreading branches and the roof of leaves. There was a damp, rich, organic smell in the air.

Deep in the woods we came to a lane, "Lovers' Lane," my mother called it. We sat beside it, hidden in the trees where we could look up into the band of open sky above the lane, and waited. My father and Oscar talked—about farming, probably, this new business my father was just getting into. Then: "*Quiet now.* I hear'm coming."

I strained to hear their familiar caws as my father pulled back the lever on the magazine of his gun, then let it slide forward to put a shell in the firing chamber. He crouched, pointing his gun at the sky.

Then: *CAW!* Sharp and alarmed. Then a rush of calls, *CAW! CAW! CAW! CAW!* with the birds' dark shapes above the trees, their whistling wingbeats and wheeling flight, and my father raising his prized twelve-gauge Browning semi-automatic shotgun and BANG, a crow crashed through the leafy branches, then BANG BANG, and BANG, and another and then another crow broke through the leaves overhead and bounced when they hit the ground. I saw the smoking shells eject from the gun and my ears rang and I heard the spent shot falling back through the leaves like sprinkling rain.

I ran to where a crow had fallen. It lay crumpled among the dead leaves with its head up, following me with a furious eye. Its beak was open. It was panting.

"Watch out for the beak!" my father called.

I reached down and the crow clamped its beak over one of my fingers. "Ouuuu!" I cried, the crow hanging from my finger. "*Ouuuuuuu!*"

My father strode up, squeezed the crow's neck and its beak opened. I pulled my finger away. Then my father grasped the crow's head and twirled its body until the body flew from the head and dropped to the ground where it jumped and flapped just like a chicken with its head cut off.

Oscar came over, swinging a crow by its legs. "Here, feel this one, Ross. It's dead for sure."

The dead crow was floppy, still warm. I held the wild, lifeless, now harmless thing in my hands and tried to fathom its strangeness, its mystery. Blood dripped from its open beak. You could see its pointed little tongue.

"Crows're smart, you know," Oscar said. "You catch a young one and slit its tongue, it'll learn to talk."

Did he pull out a wing feather and stick it in my hair and say I was like an Indian now?

And was that the start of my fascination with the woods and wild things, including Indians?

I was standing in the driveway one day where it came out of the woods, watching my father and Oscar pulling in with a load of hay. My father was driving the horses. He looked down at me from on top of the load. "Wanna come up?"

"Yuh!"

"Whoa!" my father told the horses.

I knew about horses by then; knew to walk unhurriedly behind their stamping hooves so as not to startle them, and knew to step onto the wagon tongue, then climb the ladder on the front of the wagon rack to where my father could pull me up. Then I was high above the driveway. My father went *chek-chek* against the inside of his cheek, snapped the reins, and the horses lurched into motion; the wagon bumped and swayed and the hay was soft and sweet-smelling and made a rustling sound as the wagon rolled up the driveway.

"C'mon," Oscar said. "You can sit with me."

I was snug between Oscar's knees, surrounded by his warmth and his hired man's smells and the overpowering fragrance of the long-stemmed hay, when we pulled into the yard. Joyce stood looking up at us. I laughed at the pouty, envious look on her face.

We pulled across the yard and under the overhang of the barn so that I could look into the open, second-storey mow. My father said, "Whoa. I said WHOA!" and the horses stopped. He turned to me. "You think you can slide off?"

Then I was lifted up and eased over the side of the load. I couldn't see the ground, didn't know how far down it was. My father said, "Ready?" and let me go. I slid softly down the hay to the hard ground without falling.

"You okay?" my father called from out of sight on top of the load.

"Yup!" I felt hot and proud and wanted to do it again.

Joyce came running across the yard. "That fun?"

She could never do that, I thought. My sister was a girl.

We watched the unloading. The wagon was positioned under the overhanging roof and the horses were unhitched from it and then hitched to the rope leading to the block and tackle apparatus connected to the hayfork. My father, up on the load, plunged the big, two-tined fork into the hay and "set" it, that is worked the lever that opened the barbs in the fork that would "hook" a bunch of hay when the fork was lifted, while Oscar, on the ground, eased the horses ahead until the rope's slack was taken up. "Giddup!" he said. The horses strained, the rope creaked, and half the load, it seemed, lifted to the protruding peak of the barn, snapped into the switch, then swayed along the track into the mow.

"Whoa!" Oscar hollered. My father jerked the release rope. The hay made a little shock wave when it dropped into the mow.

Oscar backed the horses now and my father gave a yank, pulling the empty fork back along the track until it clunked into the switch, then dropped—its tines just missing my father—to stick in the hay. That, and then the fork heavy with hay again, lifting toward the peak of the barn, the straining rope and the straining, farting horses kicking up dirt as their hooves dug into the ground for purchase, the barn creaking with the weight of the hay suspended under its roof until its release again into the mow and my father standing there, in the midst of that seeming violence and danger and strenuous motion, part of it, absorbed and unperturbed. Joyce stared. I stared and thought, *That's my daddy. He ain't afraid.*

Watching the milking, I saw the cows in two rows, their rear ends lining the central walk, with Oscar on one side and my father on the other, both pulling and squeezing milk into their pails. I liked the sound, familiar to me after our year on the grandfolks' farm, that soft, rhythmic *pish pish pish pish* in the fresh early mornings and warm summer evenings. Alternately, Oscar or my father would stand up, come out from between the cows, and carry his pail to the milk house. The milk was poured through a strainer into an eight-gallon milk can. When the can was full, its cover was clapped on and the can lifted over the edge of the cooling tank and settled into the cold water. Then, with a fresh filter in it, the strainer was placed over an empty can and the old filter, dripping froth, was tossed to the waiting cats. The cat that got the cloth filter swallowed it and later pooped out a turd like a cloth cigar.

Evenings especially, I hung around the barn during the milking. It was the place for me to be after supper, just as the house was the place for Joyce to be, where she could watch our mother at *her* work.

I was allowed the run of the barn so long as I didn't actually run. *Don't scare the cows.* I was getting used to the cows. I could walk down a manger, along the row of munching heads, and though a cow might suddenly thrust her nose at me and blow, I wasn't scared. You got to know how far a cow could reach in her stanchion.

Oscar showed me how a "bindle-stiff" carried things—his clothes, cans of food, his valuables. He took out his bandanna, tied the corners together to make a little sack, then poked a stick under the knots and put the stick on my shoulder with the bindle, or bundle, hanging behind me.

"There. Now you're a bindle-stiff."

After that I'd pretend to be one. I'd tie a handkerchief to a stick and fill it with an orange or apple, maybe a slice of buttered bread, whatever my indulgent mother chose to give me. Then, bindle on my shoulder, I'd tramp up the field road toward the back of the farm, to a clump of trees that became my "hobo jungle," and sit there in the shade and eat my little hobo's lunch.

Afterwards I'd shoulder my empty bindle and walk home. My mother thought that was "cute." Oscar's eyes would twinkle. But my father only nodded in distracted recognition of my imaginative development, his mind on other things.

One day my mother said, "How'd you like to walk to your aunt Betty's?"

Aunt Betty, the next oldest of my mother's four sisters, and Uncle Gerry lived nearby on the farm he'd inherited from his folks, retired now in California. When we visited there, Aunt Betty gave me hugs and made a lot of me. She called me her "favorite."

I had no idea where their farm was in relation to ours, but my mother said I could cut across the fields to get to it. It wasn't far.

She made me a lunch, a sandwich of some kind, an apple or orange, and wrapped it in wax paper. She added a glass jar of our unpasteurized milk, then helped me put everything into my "bindle."

"Now don't get lost," she said, as I set off on my adventure.

I walked up our field road to the farm dump, which was over a rise or two and out of sight of the buildings but hardly a quarter mile away, and stopped to eat my lunch. The dump was off to one side of the field road and spilled over into a ravine. Beyond the ravine was a corn field that, bordering on our land, turned out to be my uncle's.

The dump was a rock pile, rocks picked out of the fields, season after season; along with worn rubber tires, discarded license plates, scrap lumber bristling with rusty nails, old bottles and old tin cans, some rusted almost to soil—layer upon layer of such junk, like a midden on some ancient village site awaiting its archeological dig.

I sat on a rock, eating my lunch. It was pleasant and shady under the trees growing out of the rubble; they swayed and sighed in the summer wind. The field in front of the ravine rose toward a line of woods that disappeared over the curve of a hill. I might have looked up and seen a redtail hawk, turning and tilting in the thermals.

Then I found myself looking at a woodchuck, flattened against a rock. We watched each other. When I moved to take a bite of my sandwich, it scrambled off the rock and disappeared.

I finished my lunch and, with orders to save it, put the empty milk jar in my bindle. I started off again.

I walked to the fence at the end of our field road, crossed it, and found another field road, the one my mother had said would take me to Aunt Betty's. It was just like the one I'd left, two dusty tracks with grass growing between them. I walked along the road and presently saw the back of buildings I recognized as my aunt and uncle's, their barn and silo and windmill before the trees in their yard. I couldn't see the house until I was past the barn. I walked to the house, climbed the back steps, and knocked on the screen door.

"Well. Look who's here! Where'd you come from?"

Aunt Betty was only nineteen then and recently married. You could see she was my mother's sister. She took me inside and gave me a glass of Watkins nectar. When I'd finished drinking it, I was ready to start back.

"Going so soon?" Aunt Betty said. "Well, thanks for visiting, little man."

The walk home seemed to take a lot longer than the hike out. Finally I reached the dump and the rock where—ages ago, it seemed—I'd sat eating my lunch. There was no sign of the woodchuck.

Past the dump I saw the top of our windmill and the tops of the trees around the house. Then the road dipped and they were out of sight until, coming over the rise by the patch of woods where I played hobo, I stood even with the little platform under the blades and tail of the windmill and looked down on the familiar buildings of our farm. They were all still there, just as I'd left them.

When I came into the yard, my mother was hanging clothes on the line and said pleasantly, "Home already?"

That surprised me. I'd been gone a long, long time, I thought, and been far, far away.

Toward fall my mother took me to the one-room school on the way to Hamel. There was a woman inside, sitting behind a desk. There were other mothers there, too, and other kids, and we all must have waited our turn to talk to the woman, who would be my teacher.

"No, I don't think he's ready," she told my mother. "They've found that children do better in school when they start a little older. Five years old. That's pretty young."

"He'll be six at the end of December."

"That's half into the school year. No, I think he'd better wait another year."

So I was free for another year.

Monday, November 11, 1940. Armistice Day.

It had been a dry, mild fall, after a dry summer. The weekend turned cold, with misty rain. Then on Monday it began to sleet, and then to snow. The temperature dropped from above freezing to near zero Fahrenheit. Winds of thirty to sixty miles per hour began to blow. It was the start of the famous Armistice Day Blizzard of 1940.

The screens were still on the windows, my mother remembered. *Your dad had to get out there in the blowing snow and take the screens off and put the storm windows on. Typical. He always put off work around the house.*

The cows must have been in pasture. But they would have come in by themselves and huddled in the yard until Oscar or my father let them into the barn.

There was a stack of wild hay, "horse" hay, piled beside the driveway, that would be hard to get at if we were snowed in. With horses and wagon, Oscar and my father drove to the stack, loaded it onto the wagon, and started for the barn. But the horses, struggling to pull the wagon in the snow, slipped and fell, tangling their harness. *I could hear your father swearing from inside the house!* Eventually the horses were calmed, re-harnessed, and the wagon driven under the overhang of the barn and a tarp put over it.

The storm was a big excitement for Joyce and me. The wind moaned and it was blinding white outside. There was a cosy sense of shelter in the house, though cold drafts swept in under the doors and through the windows. Our mother stuffed rags under the doors to keep the snow out. Oscar and my dad went out into the storm to do the milking and almost lost their way getting back to the house after chores. No electricity (did we even have it yet?). Only the intimate, shadowy light of candles and kerosene lamps.

In a hospital in Minneapolis that day Aunt Betty's first baby, my cousin Judy, was born. Had my aunt not gone into the hospital the day before, my cousin might have been born at home, without even the help of a neighbor woman, as the storm raged outside.

Was there coal for the furnace, or had my father procrastinated there too? We must have had wood to burn anyway. The house, like most farmhouses then, was uninsulated, and we kids were put to bed early to keep warm. I wanted to stay up, of course. I didn't want to miss anything.

The next day it was still snowing and blowing. The men came in from morning chores and we all holed up inside the house. Periodically my father or Oscar went down in the cellar to stoke the furnace. We kids were bundled up because of the drafts. My mother got Oscar to play cards with her. It was strange having the men in the house all day.

At noon they went out to feed the cows in the barn. You knew they were back when you heard them stomping their overshoes outside the kitchen door. After supper they went out again to clean the barn and do the milking. Maybe I went with them then, struggling through the drifts in the yard. It had stopped snowing, I think, and the wind was down.

On the third morning we woke up, looked across the snowy willow swamp to the Old Rockford Road and the Huar farm and saw the top of a panel truck sticking up out of the drifts. It was one of the bread trucks that serviced country towns outside Minneapolis. "I'm going over there," my father said.

After the milking, he put a bridle on one of our draft horses, mounted up, and rode over to the Huar place. We watched him through the kitchen window. In a while he came back, plunging through the snow with two full gunny sacks across the horse's withers and a big grin on his face as he swept by us under the window.

He'd brought us a treasure of baked goods: donuts, longjohns, bismarks, jelly rolls, cream puffs, and bread, of course (white, I'm sure). All frozen. "The guy said just to help myself," Dad told us. "He's staying at Huar's till the plow comes through."

We all had sweets for breakfast.

Then my dad and Oscar had to get the milk out. There'd been four milkings since the storm, and no milkman to pick up the cans, and by now every milk can and probably every jar and bottle on the place was filled.

They loaded the bobsled and hitched the horses to it, then started out of the yard for the creamery, horses and sled breaking through the drifts and disappearing into the woods toward the road. They were headed for the Dutch Mill Creamery in Loretto, three and a half miles away. It would be almost dark when they got back, trotting the horses over the trail they'd made that morning, pulling out of the woods and into the yard with their load now of empty cans.

It was a week before all the roads were plowed. Oscar and my father got out the big, two-handled scoop that could be hitched to a horse and cleared the driveway so the milkman could get through to us again.

It's still the worst snowstorm in Minnesota history. There had been a record two-day snowfall of between 16.2 and 26.6 inches (compacted snow: readings were taken every twenty-four hours then, rather than hourly, as they are today) that knocked telephone and power lines down and piled drifts on the roads. Some fifty-nine people, exposed to the freezing wind and blinding snow, died in the storm (one hundred and forty-four nationally as the storm blew from the Rocky Mountains to the Eastern Seaboard). Many of the Minnesota dead were duck hunters, caught on the Mississippi River bottoms in the southeastern part of the state.

It was the start of our first winter on the farm.

Pioneer stock

My mother was a farm girl, but she never wanted to be a farm wife. She was the first of her parents' five daughters, born January 23, 1912, at home, and christened Marie Louise Ruth Hamel, in the Hamel church. She was an only child until my aunt Betty was born nine years later, followed, in regular succession, by my aunts Lois, Clarice, and Bonnie.

The Hamel farm stood, like the DesLaurier farm down the street, right in town, as it might have in Quebec or in France itself. Below the town's bluff, across the Soo Line tracks and the new Highway 55, lay my grandparents' forty acres of field and pasture. An adjoining twenty acres belonged to my uncle Lucien, Grandpa's youngest brother and the town's postmaster.

It was all that remained of an entire section, six hudred and forty acres, that had stretched north of town and that William (originally Guillaume) Hamel, born in Quebec in 1849 and brought by his parents to Minnesota Territory at the age of six, had accumulated by the time of his retirement. About that time, he gave farmland to each of his five sons. The home place (about a hundred and twenty acres) went to my grandparents, who had agreed to look after him. Yet none of the Hamel boys, including my grandfather, was the farmer old William had been. They had neither his energy nor his ambition, it seems, and wound up "pissing their land away," in the words of my energetic and ambitious father.

The Hamel pasture, with upland fields on either side of it, was mostly bottomland along the course of Elm Creek. My grandparents' cattle reached it through a concrete tunnel under the railroad tracks and the highway built, I think, as part of the construction of Highway 55. Named for the big elms that once lined its banks (since lost to Dutch Elm disease), the creek reduced to a trickle and scattered pools after spring runoff, but it always filled again after a heavy rain and, anyway, was a reliable source of water for cattle. The creek ran northeast to the Mississippi and when full there were fish in it, perch and bullheads or small catfish that had travelled the fifteen or twenty miles upstream from the great river and that as a child I tried sometimes, without success, to catch.

Downstream, past the Hamel pasture and across a neighbor's pasture, you entered a dim enclosure of bottomland forest, a silent remnant of what the early settlers knew as the Big Woods.

It was this once thickly forested region of south-central Minnesota, including its groves of sugar maples, that attracted William's father, L'Ange Hamel, after his arrival in Minnesota Territory. L'Ange, born in 1812 at L'Isle Verte on the lower St. Lawrence, had farmed there and later around nearby St. Eloi. In 1837 he married Eugenie Moffett, the daughter of a local surveyor, and together they had seven children. But it was hardscrabble living for them in Quebec, and when word reached them of the opening up of Minnesota Territory for settlement (following the hapless ceding away of their lands by the native Sioux and Chippewa), the forty-three-year-old L'Ange, perhaps with his younger brother Marcel, set out for Minnesota in the spring of 1855.

To get there they probably traveled by boat up the St. Lawrence to Montreal, by stagecoach to Buffalo, New York, by train to Chicago, and from there, by whatever conveyance, to Rock Island or Galena, Illinois, where they boarded a Mississippi steamboat for St. Paul, the territorial capital. Theirs was one of a hundred and nine steamboat arrivals in St. Paul that year, before Lake Pepin, formed by a natural impoundment some fifty miles south of St. Paul, froze in early winter and the raw little city of some ten thousand inhabitants became more or less isolated until spring. Steamboat arrivals would double and triple in the next three years as settlers streamed into the territory. Statehood came in 1858.

Upriver some eighteen miles from St. Paul was St. Anthony Falls—the head of steamboat navigation on the Mississippi—and the sawmill town of St. Anthony. On the west bank of the river was upstart Minneapolis, which would grow to incorporate St. Anthony in 1872.

It's possible that L'Ange and Marcel went first to Little Canada, the French-Canadian settlement outside of St. Paul. Whatever the case, and probably after a visit to the land office in Minneapolis to find that most of the parcels along the rivers in the area, the Mississippi and its tributary the Minnesota (then called the St. Peter), had been claimed already, they tramped northwest out of Minneapolis to the Big Woods and came to the stream eventually called Elm Creek, along which grew a wealth of maple trees whose sap could be tapped each spring and boiled down for a cash crop of syrup or sugar, as they had done back in Quebec. Here the brothers "squatted": laid preemptive claim to a quarter section or so of land by making a clearing and building a cabin before sending for their families.

L'Ange (spelled Lange by then) died in 1887, the same year my maternal grandfather was born. It was also the year that the Minneapolis, St. Paul and Sault Ste. Marie Railroad (the Soo Line) cut through the Hamel land below the settlement known then as Medina (the name of the township), constructed a depot, and named it Hamel. By that time hay was the family's principal crop, and William Hamel and his sons were hauling it off the farm to sell to owners of horse-drawn vehicles in Minneapolis—to the streetcar company (until its cars were electrified), creameries and ice companies, and private owners of horses and buggies.

Maple sugaring, too, continued on the Hamel farm. It was a spring activity until well into my mother's girlhood and had its gleeful moments, according to my grandfather.

The relatives used to come out from town [Minneapolis] *when we were sugaring. We'd make sundaes for them—you know, hot maple syrup poured over a dish of snow? They'd eat that until they got the shits.* (Wheezing chuckle.) *They'd be out there in the woods, the women, you know, in their long dresses, squatting in the snow behind the bushes.* (Cackling laugh, degenerating into a choking smoker's cough.)

But the sugaring ended on the Hamel farm not long after William died, in 1919, and my grandfather sold the maple trees for lumber. That gave him the money to buy a new car, the 1925 Dodge my mother learned to drive while in high school.

Christ, was my father's comment on that move. *The quick buck!*

There was more squandering of my grandfather's inheritance—bits and pieces of the farm sold off for ready cash until, by 1939, when we moved there, only the forty acres were left that eventually my folks rented and later bought.

Your grandpa didn't know how to manage. Not to mention his breaking off work occasionally to sneak across the street for a drink and a quick card game. He'd learn, after he quit farming (soon after my father started), that working for the Hennepin Country Highway Department—he became a steady, conscientious employee—was a whole lot easier than being self-employed, and self-motivated, as a farmer. *He was no farmer.*

My father was born in Moose Jaw, Saskatchewan, October 16, 1910, to Americans who lived in Canada. His father, William (Will) Klatte, born and raised in Lima, Ohio, was a locomotive engineer for the Canadian Pacific Railway.

My father's mother, originally Mary Agnes Williams, born in western New York, was the daughter of George Ross Williams and the former Emmer (or Emma) Wellman, pioneers of the period between the end of the Civil War and the start of the twentieth century. According to family legend, George Williams made and lost two fortunes in his long lifetime.

Born in 1849, the same year as William Hamel, he was of Welsh and some French descent; Emmer was straight English. His parents, it's said, were Quakers.

The old pioneer came from New York State's Otsego County, near Cooperstown, the setting for a couple of James Fenimore Cooper's Leatherstocking Tales and the future site of the Baseball Hall of Fame. In 1863, at age fourteen, he entered the Union army as a drummer boy.

He went to the Civil War in place of an older cousin who had been drafted under Lincoln's conscription law of 1863 and who, so the story goes, failed to pay him as his substitute. (In the North, you could avoid the draft by hiring a substitute or paying the federal government a fee of three hundred dollars. It was the draft and this fee, which favored the rich, that led to the murderous Draft Riots of 1863 in New York City.)

    We used to hear of these and other old family grievances from my great-aunt Lelah, Great-Grandpa Williams's oldest daughter, during visits to our farm. World War Two was going on, but Aunt Lelah, having come out to the farm with my aunt Stella, my dad's only sister, and her husband Rex, would ignore all talk of the current war to rail on about The War Between the States and her father's heroic and unrecognized part in it. She must have been pleased when wartime conscription began in 1940 without—"officially, anyhow"; Aunt Lelah was nothing if not skeptical—the comfortable alternative of substitution or commutation that had put her father in harm's way when he was only a boy and allowed his cowardly and dishonest cousin to remain safely at home.

    As a young woman Aunt Lelah, like Laura Ingalls, future author, as Laura Ingalls Wilder, of the Little House books—and probably about the same time—taught school on the Dakota frontier, and in fact taught her younger sister Agnes, my paternal grandmother. I used to think of Aunt Lelah as an old maid until I learned that she too, like my father's mother, was a widow.

    I knew her as the old fuss-budget who came out to our farm, veiled against the elements, and allowed us kids, doubtless pretty grubby looking, to peck her cheek—*through her veil*. We used to laugh about that behind her back. Still, she was a lively old bird, sharp as a tack, and loudly and amusingly opinionated in the Klatte-cum-Williams manner.

Two or three of Aunt Lelah's old steamer trunks wound up being stored in our upstairs "junk room" on the farm. At some point, I managed to jimmy them open, and so discovered their treasures.

Among piles of folded, mothball-pungent clothes, pressed gray flowers, ancient locks of hair, old greeting cards, old invitations, old photographs and yellowed letters, I found a toy tugboat, all metal and made to scale, with a windup motor and movable rudder. I ran it in circles in the milk tank and on meltwater ponds in the spring. And there were interesting old books: late nineteenth- or early twentieth-century editions of *The Three Musketeers* and *Twenty Years After* by Alexander Dumas, *The Harbormaster* by Theodore Goodrich Roberts, *Camp and Trail* by Stewart Edward White (full of still-useful information), a two-volume little pocket edition of *Robinson Crusoe*, and a wonderful old set of literary anthologies called, I think, *The Golden Books,* and whose individual volumes, bearing titles like *The Golden Book of Adventure, The Golden Book of Romance,* etc., Joyce and I read and reread, stretched out on blankets on our lawn, through many a summer afternoon.

The most historically interesting item I found in those old trunks was an ornate little autograph book, inscribed with period sentiments, from when Aunt Lelah was a schoolgirl in the 1880s.

> *Preserve this page in memory of*
> *one who wishes you a happy and successfull*
> *journey thro' life.*
> Your Uncle

Chas. S. McLeury
Sheldon, Dec 8 1882   State of Iowa

A Token of Love from
Your uncle,
Geo. O. Wellman
Sheldon
Dec. 9th 1882   O'Brien Co., Ia.

Sheldon was on the prairie of northwest Iowa, to which the Williams moved from upstate New York about the time of the first autographs in Aunt Lelah's little book. Ten years before, February 6, 1872, in Milford, N.Y., George Williams, by then a carpenter, and Emmer Wellman, had married. They had four or five children before their move to Iowa.

One of Emmer's brothers, Charles Wellman, had gone first to Iowa, and a sister of George's had lived there with the Wellmans before marrying Charles McLeury, a banker, who later built what became known as the McLeury mansion in Sheldon.

Later entries in her autograph book indicate that Aunt Lelah (or Lepha, as it's sometimes spelled) went back East, probably to attend school:

Lelah.
Let not our friendship
Like the rose wither,
But like the evergreen
Last forever.
Allie Wilber
Alpha Society

Mt. Vision, N.Y.
Oct 28th 1886

About this time the Williamses moved from Sheldon to Sisseton, Dakota Territory (in present day South Dakota), where they staked a claim outside town and my great-grandfather started a "dray" line to haul mail and supplies between the Soo Line railhead at Big Stone City and the Sioux reservation at Sisseton. As a delegate to the South Dakota state convention, he would have been in Pierre (pronounced Peer) in the fall of 1889, the year the Dakota Territory was divided into the states of North and South Dakota.

A favorite story of Great-Grandpa Williams's was about the time he was crossing a creek in spring flood with a load of freight when the wagon bumped over a rock under the water and, without realizing it, he lost a sack of dried prunes ordered by a neighbor. When he discovered the loss, he drove back to the creek and then into it, struck the same rock, felt around in the swirling water and pulled up the missing prunes. He took them home, washed and spread them out to dry, and had two sacks of prunes. He took only one to his neighbor, kept the other for himself, and had the man tell him later: "By Golly, those were the best prunes! I'd like to order some more!"

His wife's "going off" was another story, probably *not* a favorite. She was alone with her small children out on the Dakota prairie (her husband off politicking or hauling freight or working their claim or constructing buildings in Sisseton) when there was a storm one day, perhaps a tornado, and she was struck by a flying board. In the torrential rain that followed, the Williams cabin was flooded. (My grandmother remembered being lifted above the rising water and put on a table by her mother.) Was it the board or the flood or just her fearful loneliness that left Emmer Williams "not right" afterwards? In any case, she stayed behind under Aunt Lelah's care when Great-Grandpa Williams, following the latest boom, went north to Saskatchewan. She died in Sisseton. Her husband stayed in Moose Jaw.

My grandmother, meanwhile, had grown up and gone to live with her older sister Estella and Estella's much older husband in Minneapolis. "The Major," as he's sardonically called in my family—supposedly his rank in the Union army—had cut quite a figure around Sisseton with his buggy and matched pair. He courted Estella, said to have been the most beautiful of the Williams girls, and despite his age (he had an eighteen-year-old son, Estella's age then, by a previous marriage) was considered a "catch" by her parents.

She was sacrificed. To a man older than her own father who had everybody fooled until he showed himself for what he was.

That's my aunt Stella, full name Estella (named after *her* aunt), and she's referring to the Major's flaws of character, among them his swindling of a pile of money from my great-grandfather Williams (presumably one of his lost fortunes).

Grandpa was gullible. So was my mother. It's a Williams trait.

And my uncle George on "the Major":

He made Estella's life a hell on earth and tried to practically rape my mother.

That caused young Agnes to leave her sister's house and board with another family in Minneapolis. She'd attended Emerson High School and was working in Dayton's department store when her father came down from Canada and talked her into keeping house for him in Moose Jaw. The prospect may have excited her. As it happened, she met Will Klatte up there.

His full name was William Henry Klatte, and he must have been a romantic figure to young Mary Agnes Williams with his blond, Teutonic good looks and the fact that he was a CPR engineer. The railroad engineer then was something like a prince of the working class. Up there in his great engine, leaning out the window to watch the tracks ahead of him, he was the captain of a dry-land ship, the envy of men and boys and perhaps a sex object to women. Among American culture heroes, you could place him somewhere between an 1850s river boat pilot and a 1920s aviator.

Agnes Williams, judging from her early pictures, had a gypsy look. She's said to have been a tomboy on the Dakota frontier, liked to play with Indian kids from the reservation, and earned a nickname she didn't much like: "Little Indian."

Will, the son of German immigrants, had come up to Saskatchewan from Ohio in response to the advertised need for railroad engineers on the Canadian prairies. He held an engineer's license, but because of his youth and lack of seniority his chances of driving an engine in the States before middle age were remote. In Canada, he soon found himself at the controls of a CPR locomotive.

The two met in the Moose Jaw rooming house in which the twenty-year-old Agnes and her middle-aged father, off his claim on the cold prairie for the winter, were staying and where Will was a fellow boarder. Will and Agnes were married, in Moose Jaw's Episcopal church, in 1906.

They would have nine years together, and three of their four children were born in Canada: George, Mervin (my father), and Estella. William (my uncle Billy) was born in Lima after Will's death.

Among the photographs I have of Will Klatte is one, taken in 1914, in which he stands on the Saskatchewan prairie holding his shotgun and with a string of dead ducks over his shoulder. In another picture, taken in front of the Klatte house in Moose Jaw, he's presenting his shotgun, butt first, to my four-year-old father, who's reaching to place his finger on the trigger. One of my father's vivid memories of his father was of the day Will took out his "unloaded" shotgun to clean it, aimed at a picture of flying ducks on the wall, and *BAM!* destroyed the picture and put a hole in the wall. *I can still see the surprised look on his face.*

Will Klatte, it's said, was a passionate hunter and a great wing shot, a passion and an ability my father inherited. His quick reflexes and wonderful hand-to-eye coordination, his instinctive skill at snap deflection shooting, allowed my father to knock flying birds out of the sky or down running deer in the woods in some seemingly effortless way that I tried fiercely to match as a boy and young man and never could. *You gotta lead'm,* my father would tell me, but it was more complicated than that. When a pheasant flushed, for example, you had to visualize where the bird was going, its speed and angle of flight, its rate of ascent or descent, then calculate how much to lead it, whether to shoot over or under it as well as ahead of it, all within a second and without apparent thought, the way your foot slams reflexively on the brake pedal when something jumps in front of your car.

Will Klatte was killed, on December 30, 1915, in a train collision outside of Moose Jaw. He was steaming his water train toward the city in a blizzard and struck a stationary freight train on the tracks ahead of him. That train had stopped, ironically enough, because it was low on water. The crew—perhaps loath to face the driving snow—hadn't gotten around yet to setting flares on the tracks when Will's locomotive came chuffing out of the storm and slammed into the train's caboose. Will was killed outright and his fireman "seriously injured." The *Lima Times-Democrat* for January 3, 1916, puts Will's age at thirty-five at the time of his death; but if the date of his birth (March 1879) is correct, he was thirty-six.

The accident was considered Will's fault and, until the truth came out about the absence of warning flares, my grandmother was denied her CPR widow's pension.

Will's body, accompanied by his pregnant widow and her father—he'd turned sixty-seven by then and was about to retire in Minneapolis—was taken to Lima for burial.

And here I have only our side of the story. There must have been two versions of it, but those of the Ohio Klattes who might have told their side are gone. Our side has my grandmother, alone now with three kids to support and a fourth on the way, being denied Will's life insurance.

Will's mother kept the money.

She was able to do this because Will had neglected, after his marriage and before his death, to change the beneficiary on his life insurance policy from his mother to his wife.

Will's folks, I guess, were in financial straits at the time, and his life insurance must have been a saving windfall. Certainly the money belonged to my grandmother by moral, if not by legal, right. Had she been more assertive (meekness was one of her virtues or a fault, depending on who you're talking to), she might have argued her case. And her case was desperate: she was a widow and single mother at a time when there was no welfare, no support at all for a woman in her situation except family, if you had one; otherwise, it was the "poor house."

Of course the destitute widow was offered a home in Lima but with the stipulation (again, this is our side of things) that her kids be raised Catholic. Will, though a lapsed Catholic during most of their marriage, had returned to the church by the time of his death—probably under pressure from his visiting sister. Agnes herself was a Presbyterian at the time of her marriage, was now a Methodist, and would eventually turn Christian Science. Raise her kids Catholic? That decided her. After Will's burial in Lima and following the birth of my uncle Billy on August 1, 1916, Agnes Klatte took herself and her non-Catholic children up to Minneapolis to live with her father.

American dreamers

My father was only two months past his fifth birthday when his father was killed. The loss of a parent during childhood, psychologists tell us, is always traumatic. Perhaps that was the origin of my father's anxious drive as an adult, his explosive temper, the sense he gave you sometimes—when hay was down, for instance, and it looked like rain—that he'd gone a little crazy.

In Minneapolis the widowed Agnes Klatte, with her four children, found her supposedly retired father still working as a building contractor and living, I'm not sure where, either in rented digs or in the small house he may have built by this time above Cedar Lake. It was the bitter cold on the Saskatchewan prairie that had started the old pioneer going "south" to Minneapolis for the winter. He worked through his seventies and built a last house after he'd turned eighty.

So his daughter kept house for him again, as she had in Canada before her marriage, in exchange for room and board. Eventually she got her small widow's pension from the CPR, but until her children were in school her only other source of income was through taking in washing and baking breads and pastries at home to sell.

Eventually she found outside work as a practical nurse (and, more often than not, as cook and housekeeper) in the homes of well-off Christian Scientists. They were people who refused regular medical care because of their faith that promotes physical, and by extension, mental health through a study of Mary Baker Eddy's *Science and Health*. The poor widow and single mother was soon converted.

In the early thirties she nursed the sickly wife of a man named Loomis who, after his wife "passed on," as Christian Scientists say, became her second husband. Old Loomis (always Mister Loomis) at first proposed an "arrangement," but Agnes Klatte insisted on marriage—a marriage of convenience that she hoped would provide a home for her then eighty-seven-year-old father as well as for herself. The year was 1936.

Her father was taken into the Loomis home, but then, sometime after Loomis himself sickened and died, (in 1941), and his "spoiled" only son, the infamous Arthur (*Arthur*, his name always spoken in my family with the same contempt as *The Major*, that other villain in our history), forced her father to leave. The old man "embarrassed" him, Arthur said, in front of his friends.

Agnes moved out of the Loomis house herself then and into an apartment in Minneapolis, while my great-grandfather Williams, after staying on with us in the Cedar Lake house and later moving out to Hamel with us, finally returned to the city where he was looked after by his daughter, now Agnes Loomis, until his death, at ninety-three, in 1942.

As for selfish Arthur, you could hear the muted satisfaction in my mother's voice when she told of what became of him. *He died young. I don't think he had a happy life.*

The short, sweet-natured old woman I knew as Grandma Loomis was at once workaday practical and religiously, one has to say, impractical. She was of pioneer stock, after all. She was also, after the death of her beloved Will, a Christian Scientist with the stubborn belief that all unpleasantness, all pain, even death, is an illusion. She followed her faith to the extent of thinking it unnecessary to take my father, as a child, to a doctor after he'd broken an arm; a neighbor, fortunately, intervened.

And yet, in many ways, she was more open-minded and considerably less prudish than my repressed Catholic mother. Example: I have a copy of Karl de Schweinitz's *Growing Up*, his little book first published in 1928, which instructs in what parents used to refer to as "the birds and the bees." I have it in the 1944 revised second edition, inscribed in Grandma Loomis's old-fashioned hand: *May 20, 1945. To the Klatte children. From Grandma.* She meant it, I think, to be read to or by us kids, but leafing through it my mother found the pictures a little too graphic, and came upon such passages as the following:

The sperms of men, like those of the four-legged animals, live in two testicles in a little bag under the penis. The father places the sperms in the body of the mother in very much the same way that the four-legged animals do, only the mother and father can lie together facing each other. The penis then fits into the vagina of the mother which has its own opening underneath the opening for the urine or waste water.

That was too much for my mother, farm-bred though she was. She was especially shocked, I think, by the comparison of human coupling to that of animals.

My father, though small and skinny, was athletic as a kid, playing baseball, football or hockey, in and out of school. He also, with his older brother George, enjoyed nature such as it existed in and outside of south Minneapolis, roaming the shores of Cedar Lake or Lake of the Isles and exploring west along the Burlington Northern railroad tracks through the suburbs of St. Louis Park and Minnetonka or along the wooded course of Minnehaha Creek.

He had a taste of glory at age fourteen, in January 1925, when he won the annual dog sled race during the St. Paul Winter Carnival. Contestants, all kids on coaster sleds, were yanked around the two-mile circuit of Lake Como by single dogs. My father's dog was a big mongrel collie he and George found as a stray puppy and trained to pull a sled. The trick was to stay on your sled and keep your animal from fighting with the other dogs as you raced with the pack around the frozen lake, bellyflopped, ice and snow pelting your face. There's a newspaper photo of my father, taken after his victory, seated on his sled in a knit cap, wool sweater and high, laced boots, posed opposite Skippy, his champion dog. He looks familiar somehow, thin and handsome, a boy hero. Then you notice his resemblance to the boyish-looking Charles Lindbergh of Little Falls, Minnesota, before he grew up to make his famous flight across the Atlantic.

About this time, like some character out of a Scott Fitzgerald short story, my father started caddying on golf courses, learning "the gentleman's game" from the bottom up. This was during the Boom of the 1920s, on the links of country clubs where he was dazzled and sometimes humiliated by the golfers he served—and by some of the other caddies, snotty sons of the privileged—a poor boy among the rich. Not surprisingly, he dropped out of high school after only a year, too anxious to *get going* to sit still in a classroom.

Meanwhile, he'd became the favorite caddy of a young man hardly older than he was, *a guy who didn't have to work*, whom my father always spoke of as *Mister* Weatherby. Weatherby, living off a trust fund, I guess, played golf more or less full time and, with my father lugging his bag of clubs, competed in amateur and semi-pro tournaments around the Upper Midwest. Together they watched Bobby Jones play in the St. Paul Open. My father played golf too, of course, once taking the lead and drawing a crowd in a tournament at Alexandria, Minnesota. But then: *I blew up under the pressure.*

He learned to box and, prize fighting being a road to fame and fortune more common than golf for the underprivileged, dreamed of becoming a Golden Gloves champion, then turning pro. But a couple of amateur bouts and the sight of all those punch-drunk has-beens or would-be contenders hopefully working out in a downtown gym discouraged him.

Too old for caddying now and with the 1920s Boom become the Great Depression of the 1930s, he took any job he could find, never surrendering, meanwhile, his American dream of *amounting to something*—making money, in other words.

Inevitably he continued working for the rich, mowing their massive lawns, polishing their boats, cleaning out their swimming pools. Once he helped build a clay tennis court on a Wayzata estate only to be fired on some pretext by the contractor just as the job was about finished and he was expecting to be paid. *I was a green kid. Well, I learned my lesson.* That, and similar experiences, caused him always to identify with "the little guy," goaded him to become his own boss, and brought him finally, through native intelligence, luck, and stubborn ambition, to within hailing distance of the "rich bugs" he as much hated as admired.

At age twenty or so, after working for a time with his brother George on a farm near White Bear, north of St. Paul, he was offered a job on a poultry farm outside Wayzata by the owner, a man he'd once caddied for. Something happened there, some altercation with a bullying fellow worker that had my father, in a sick rage, going up to his room above a barn for his Browning semi-automatic shotgun. He was jamming shells into the magazine, intent on blowing the man's head off, when the owner of the place caught up with him, calmed him down, and so saved my father, according to my uncle George, from "going wrong." A year or so later, he was made manager of the place.

Then, through the matchmaking efforts of a friend of his and a cousin of my mother's, he found himself paired on a blind date with *one of those little French girls from Hamel.*

Sitting in the back of his friend's car and necking with my mother on their first date, my father kept opening the window to poke his head out. "What's the matter?" she asked. "I get carsick," he answered. But my father was gulping air for another reason. *Your ma was so beautiful I couldn't breathe. I had to stick my head out the window to catch my breath!*

The year was 1931. My mother had been going steady with a Hamel boy since before her graduation from Wayzata High School in 1929. (She was the school May Queen that year.) He was handsome, and appropriately French and Catholic; they'd grown up together, and in old photographs they look made for each other. But she grew bored with him, their "dates" consisting mostly of being "dragged" by him to a local baseball game every Sunday.

Eventually that led to the blind date with my father.

Your dad was full of life. He was a working fool!

His ability to work, in fact, won the approval of his future mother-in-law to the extent that she could tell him, "Mervin, we'd have made a pair!"

He could throw childish tantrums, though, when things weren't going right, because he was never all that sure of himself. *I didn't have the education. I wasn't born with a silver spoon in my mouth.* So he was driven. And he learned things the hard way, the only way he knew how.

He and my mother were engaged for two years, saving for their nest egg. Finally, in the first of many such ventures, my father bought a brood of baby turkeys from his boss, "at cost," and my mother raised them on the Hamel farm. There's a snapshot of her, taken the summer before their marriage. She's in the turkey yard, tanned and windblown, squinting into the sun, holding a scrawny, immature bird in her arms and looking like the heroine in a Willa Cather novel.

That fall, after butchering and selling the turkeys and buying a car, *a cute little Model A coupe with a rumble seat,* they were married, November 30, 1933, in Hamel, the first couple to be joined in the newly built St. Anne's Catholic Church. My mother was approaching twenty-two, my father just twenty-three.

In a wedding snapshot, taken in front of the screened porch of the Hamel house, my mother, in a long dark dress—*blue velvet, store-bought. We couldn't afford white*—and a pillbox hat cocked stylishly over one ear, stands with her veil draped over an arm and holding a big bunch of chrysanthemums. She's smiling gorgeously into the camera. Beside her my father, in silk tie and tightly tailored suit with a carnation in the lapel, bears a wolfish grin. She looks like a movie star. He looks lean and hungry.

Your father weighed only a hundred and twenty pounds in his overcoat when we were married. I weighed more than he did!

That first summer after their marriage, the drought and Depression summer of 1934, the young couple lived in a rented cottage on Cedar Lake. They'd moved there that spring, from a one-room, third-floor walkup off Lake Street, after my mother got up one morning, smelled the chlorinated tap water in the bathroom down the hall, and threw up. *That's when I knew I was pregnant.* She was a country girl, after all, and hated city water. The Cedar Lake cottage had a well.

On the hill above the lake my paternal grandmother, not yet married to Mister Loomis, still lived with her father, old George Williams, in the house he'd built on Drew Avenue. Every Monday, "wash day," she'd walk down to the cottage for my mother's laundry. Despite Ma's protests, she'd collect it, pack it up the hill, then return with it that evening, all washed, ironed and folded.

It was the hottest and driest summer anyone could remember. The air was often hazy with dust, topsoil drifting eastward from the "black blizzards" out in the Dakotas, and in the city there was periodic violence—the bloody Truckers' Strike in Minneapolis that had started in May and wouldn't end until October. It was a fearful and yet a hopeful time, a time of waiting: waiting for the Depression to end, waiting for rain; waiting, in my mother's case, for the birth of her first child.

As a Catholic she'd been more worried about not getting pregnant right away than about the hard times she and my father were starting out in. My father had a job when a lot of people didn't—two of them, in fact, part-time in rival filling stations—and so they were doing all right. *Then I got stupid.* Eager for a commission, he sold a set of tires at one place and, at the customer's convenience, installed them at the other. Caught in the act by his boss there, he was promptly fired and so their income was cut in half. *We were down to thirty-five bucks a month!*

And they had to move, from their "nice little efficiency" on Humbolt and Lake, their honeymoon apartment, to that single, depressing room off Lake Street with only a two-burner hotplate for cooking, where my mother spent her days reading or listening to the portable radio my father had won in a movie theater draw, occasionally going out to window shop—killing time until my father came home from work.

Now, because the Cedar Lake cottage wasn't winterized, they'd have to move again when cold weather came.

Meanwhile, they had the lake and plenty of the shade trees my mother loved. She couldn't swim and was afraid of water (my father was the swimmer; he took dips in the evenings), but down the shore in a circus tent was cheap entertainment for both of them: the continuous, brutal, fascinating spectacle of a dance marathon.

It was a spectacle born of the Depression, an invention of the times to cash in on the times, a dream of big money for the contestants.

It went on all that summer, it seemed, all day and all night, day after day and night after night, to the tired strains of repeated dance tunes that carried endlessly over the lake. You could listen to it free.

Or, for the twenty-five-cent price of a movie, you could watch the marathon itself. Often during those hot summer nights my folks walked down the shore to the big tent. Inside, the bleachers were always crowded. There was the smell of canvas, hot dogs, hot coffee. And something else, the smell of bodies, the stink of nervous exhaustion, from the dance floor.

Hawkers sang out, "Popcorn! Candy! Cold pop!" Most people couldn't afford to treat themselves and brought lunches. *Some stayed all night.*

The contestants, mostly young and otherwise unemployed, who'd managed somehow to pay the entrance fee, the best of them athletes of a kind, some entertainers of a sort, alternately "danced" and walked, energetically at first and then in a stupor, for forty-five minutes of every hour, hour after hour and day after day, for weeks—sometimes for a month or more—until the winning couple was declared. Bands came and went, relieving each other on the stand, while an announcer blared through a microphone, "talking up" the dancers, calling "novelty" events such as elimination races, especially in the early stages of the contest when the crowd of dancers needed thinning out. There were scenes of comic relief, of outrageous, awkward or pathetic playing to the spectators by the various "characters" on the floor; embarrassing losses of control, women and sometimes men weeping, telltale stains suddenly appearing on the front or back of a man's pants or a woman's dress—touching, deeply human scenes of such suffering, such tenderness or, astonishingly, such eroticism you felt as if you were watching something too personal and looked away.

So it was described by my mother and later made real for me by the 1969 movie version of Horace McCoy's 1935 novel *They Shoot Horses, Don't They?* After seeing the movie, I read the tough little novel.

My father, before he met my mother, had once "managed" a marathon team—one of his early entrepreneurial ventures. For a promised fifteen dollars a week and a share of the prize money if they won, he served a pair of hopefuls by massaging their aching feet during the breaks, doing their laundry, waking them up when it was time to go back on the floor. His couple lost, of course, and he didn't make a dime. *Aw, the woman didn't come to in time after one of the breaks.*

I can see my father then, quick and eager, moving among the slow-motion couples on the dance floor, who took turns sleeping, one partner holding the other up somehow while continuing to move (*You had ta keep moving*), and the "solos," dancers whose partners had been eliminated, shuffling weirdly by themselves until their grace period was up—or, if they were lucky, until they found another solo of the opposite sex to couple with. *I would've entered the damn thing myself, but I didn't have the fee or a partner.*

Would my mother have joined him, had they known each other then? Maybe. She eventually went along with his schemes, though they might appear reckless or foolish to her, doomed to failure, a waste of effort or money. It was my mother, in fact, who liked to tell of the dance marathon that summer. She was an avid listener of radio soap operas and loved movies, but this was something better than any soap opera, better than any movie. It was real life. It was triumph and tragedy played out before your eyes, a grim yet somehow heartening demonstration of human struggle that was at once a product and a dramatization of the Great Depression. She saw that, and she was fascinated.

What I see now, of *her* then, is my young mother not watching the marathon but listening to it in the hot dark of the Cedar Lake cottage (*Electric lights cause more heat, you know, they cost money*), sitting in a rocking chair, rocking to the music that drifted over the lake all that hot and dry summer of 1934 in Minneapolis.

Like the dreamy heroine of Theodore Dreiser's first novel, *Sister Carrie*, she sits rocking, rocking in her chair beside my father already in exhausted sleep in their bed—rocking at the start of her marriage to him, the start of the adventure of their life together—rocking in their dark cottage to the incessant music in the air that spoke of the incessant, tortuous contest going on nearby. Rocking, like Dreiser's Carrie (while I floated in her womb), dreaming what dreams?

Waiting for me to be born.

THE SECOND YEAR

Toward spring of our second year on the Pepin farm my mother noticed that Oscar wasn't well. *He looked so peak-ed. And he wouldn't see a doctor.*

He was short of breath and seemed old now. How old was he? His hair wasn't gray. It was thick and sandy-colored, Ma remembered. But he was pale, she said, his eyes sunken into his head. He looked sad.

It was his weak heart, caused by the rheumatic fever he'd had as a child, and of course he smoked too much and maybe drank too much when he went to Hamel. *He looked like death warmed over.*

The fall before, he'd come back from the state fair with a present for me. It was a balloon that, blown up, formed a toy blimp with GOOD YEAR painted on it.

There was even a little cardboard cutout that, folded and pasted to the bottom of the balloon, became the blimp's car, complete with painted-on windows and passengers' faces looking out.

To give me a treat, he launched the little blimp from the top of the windmill. It floated downwind all the way to the straw pile.

He climbed the windmill again, but this time the blimp floated to the barbed wire fence past the barnyard and popped.

"I'll get you another one," Oscar said. "Next year at the fair."

But he was dead the next year by fair time.

Sundays he'd come into Hamel with us and hang around town until ten o'clock Mass at St. Anne's was over and the saloons opened up. He'd be in one of the beer joints while we visited the grandfolks after church and maybe stayed for dinner. When it was time to go home, time for chores, my father went looking for him in Pepin's or Sheridan's or Johnny Cavanagh's.

Once, at least, I was sent in to get him, into Pepin's, which maybe says something about my folks and that period when children were allowed inside a tavern. Pepin's was loud with talk and laughter that Sunday, crowded with men and a few women of the kind my mother and my grandmother called "wild" or "tough." I stood by Oscar at the bar, a little frightened. He must have bought me a pop while he drank a last whisky or a beer. Suddenly there were shouts from the line of booths along the far wall, and two men stood up and started punching each other. They grappled to the floor. "Stay close to me," Oscar said, and I pressed against his legs and he put an arm around me. The fighters rolled on the floor like schoolboys, swearing at each other, and the crowd of drinkers stepped back to give them room. It was so interesting I wasn't scared anymore.

I'd seen one or two other fights in Pepin's, though from safely across the street on the grandfolks' lawn, when the brawlers had spilled out the batwing doors and off the veranda onto the gravel in front of the place; afterwards I looked for the change that had fallen from the men's pockets as they scuffled on the ground only to get up afterwards and go back into the tavern together, leaving their scattered dimes and nickels, quarters and half dollars, on the ground for a kid like me to pick up. Finders, keepers. I learned to do that from the town kids that year we lived in Hamel.

Johnny Cavanagh's much smaller joint was another hangout of Oscar's—and occasionally of my grandfather's. It was where Mary, his blind daughter, with a finger inside the glass that told her when the glass was full, pulled beer for the boys. When not behind the bar she sat hunched over the big heat register in the floor, picking her nose, so the joke was to ask Mary to draw you a beer—*without the finger!*

I associate Oscar's death with the smell of lilacs. That would make it early June, June of 1941. When I asked my mother where the wake was held, she couldn't remember. But I think I remember. I see Oscar laid out in the tiny living room of our house on the Pepin farm, and people gathered outside in the warmth of that summer evening, talking, taking turns going inside to stand over the open casket. It was like a family gathering, only burdened a little by the heavy smell of the lilacs in bloom in front of the house and the lifeless body inside.

That was the summer I ran away, the summer I was six. Oscar wasn't with us anymore, and the farm was quiet in the mornings after breakfast while my father slept. He still worked nights at Coca-Cola, coming home in the early mornings to milk the cows and eat breakfast before going to bed to sleep till mid-afternoon. We kids had to be quiet while Daddy slept.

That was the summer, too, we got our first Surge milkers, a pair of them. I watched my father attempting to put one of the machines on a cow for the first time, the cow kicking, my father cursing. The cow kicked him across the aisle into the hind legs of another cow, which kicked him back. He was like a ball tossed back and forth. I laughed, and my father snarled, "What's so *funny*? Go on, goddamnit, get outa here. *Move.*"

I moved when my father told me to. He could be cheerful, whistling; he'd slap my mother's fanny and she'd say, "Don't," but I could tell she liked it. Often, though, he was crabby. It was because he worked nights, my mother said. Never got enough sleep. Fell asleep when he was driving. Woke up one morning with his car in a field. Snored in the bedroom until it was time to get up to do farm work. Then he'd be pressed for time, alternately cheerful or short-tempered, depending on how things were going.

When we kids were naughty, our mother'd say, "You *mind*, or I'll tell your dad." That was enough. We minded.

So, when I broke the windmill one day and went into the house and told my mother and she said, "I'll have to tell your dad," I lost my nerve. I followed her into the bedroom and watched her shake my father awake, but as he sat up still dazed with sleep, I ran past him and out the door that opened onto the front porch and down the steps and across the farmyard. I was afraid to look back, afraid my father was right behind me, so I kept going, up the field road past the barn, running, until I reached the clump of woods where only yesterday, it seemed, I'd gone with an orange in my kerchief bindle and pretended to be a hobo. There I stopped and looked back at the farm, my lost home. I expected to see my father, outraged at his loss of sleep, charge out of the house and head up the field road after me. But no, the farm remained quiet, the yard empty. The windmill turned brokenly in the wind.

It was a windy day, but before this the windmill hadn't been turning—it had been braked. I'd watched Oscar or my father lift the wooden handle clapped to one of the metal struts to start the blades turning in the wind. I thought I could do that. I climbed a few rungs of the ladder to reach the handle and with some effort pushed it up. That released the brake and allowed the windmill's blades to respond to the wind. There was a wooden shaft running up through the fifty-foot derrick that connected the turning blades on top to the pump below. It worked like a piston rod, rising and falling to crank water up from the well into the cooling tank in the milk house, the overflow running through a pipe to the drinking cups attached to the cows' stanchions in the barn and then to the stock tank in the barnyard.

Gravity feed, my father called it. The drinking cups were a marvel, worked by the cows themselves; when a cow was thirsty it just pushed its nose into the cup, depressing the lever at the bottom, and water seeped up. One of my little jobs periodically was to clear the cups of their clogging accumulation of cud.

But now the blades of the windmill were spinning so fast and the pump was working so hard and water spilling so furiously into the milk tank, it scared me. I tried to stop it. I climbed the ladder again and pulled on the brake handle, but I couldn't close it. Desperate, I stepped off the ladder and swung my full weight from the handle. There was a *crack!* The wooden shaft snapped, and the pump abruptly stopped as the lower half of the broken shaft fell against the inside of the derrick and the upper half dangled uselessly above it while the blades of the windmill kept furiously turning. The whole wonderful mechanism had come undone, like a windup toy wound too tight.

Of course I had to tell my mother, who of course had to tell my father, whose anger could be so frightening, whose face seemed to swell sometimes into an ugly distortion of itself and whose words could strike like blows, knocking the breath out of you. I'd never gotten a licking from him. But the possibility, the threat, was sometimes there.

So that's why I ran away. I started crying, and thought of Oscar, his blues harp-sounding voice intoning, as it used to when little Marcia started crying, *There's that chin-music again.*

Past the clump of woods, past the dip in the field road, was the potato patch where a day or two before

Joyce and I had picked bugs off the plants, collecting them in coffee cans for a penny a bug. Then up past the potato field to the dump. Here's where, last summer, a lifetime ago, it seemed, I'd stopped to eat my lunch the day I walked to my aunt Betty's.

I hid in the weeds around the dump and waited for my father to appear on the field road. When he did, I would run down the bank, across Uncle Gerry's field, and into the trees on the other side. If my father kept after me, I didn't know what I'd do.

It was high summer. A hot wind sighed through the trees and sent waves across the fields. Cloud shadows moved across them. The wind and the moving clouds, the sighing trees and the waving fields, they engulfed me with a sense of the aliveness of the planet, its total unconcern, my utter loneliness.

All that day, it seemed (it was probably only an hour or two), I walked from the dump to the hill overlooking the farm, then back to the dump, over and over, wondering what to do. I'd crawl over the brow of the hill to look down on the farm. The yard was empty, the house still quiet. Didn't anyone miss me, didn't anyone care? What would I do when it got dark? I was already thirsty and getting hungry.

Then, peering over the hill, I saw my mother, with Joyce and little Marcia, come out of the house and get in the car. They were going to Hamel—without me! I jumped up and ran after them. "Wait for *me!*" I called. "Wait for meee!"

My mother had backed the car around and started down the driveway, but now she stopped. She'd seen me! I felt a rush of relief and gratitude. Then the car door opened, as if for me to climb in, and my mother looked out with a sly grin on her face. "Ohhh" I said, realizing I'd been fooled. "I *knew* that would get you," my mother said. "You forgot all about a licking, didn't you, when it came to missing a trip to Hamel."

I never got a licking. And one day, with my eleven-year-old aunt Bonnie, my mother's youngest sister, I watched my father fix the broken windmill. We sat on the ground below it, chatting away as my father stood on a crosspiece up near the top of the derrick, attaching a new wooden shaft to the rig. I guess we were sounding a little too jolly because my father growled down at me, "Yeah, well you break this windmill *again*, mister, and you *will* get a licking."

Joyce said to me afterwards,"That was a good trick we played on you, wasn't it."

"You shut up!"

I picked up a stone to throw at her, but she backed away and looked so frightened I was ashamed. I put the stone in my pocket for some reason and said, "Want me to show you where I hid?"

"Okay."

I took her up the field road, past the little woods, to the dump.

"There," I said. "That's where I hid." The weeds were still flattened where I'd lain that day like a scared wild animal, watching for my father to come after me. On impulse, I took the stone out of my pocket and placed it there.

I would look for the stone after that, after the weeds had straightened again and there was only the stone to mark the spot. It was like a piece of me, a little monument to that place where I lay one day under the windy sky and felt all alone in the world.

I think Joyce enjoyed the trick that had lured me home after I ran away because, a while before, I'd played a trick on her. It was a mean little practical joke of the kind that little boys play on little girls.

I'd found an empty perfume bottle under the front porch one day, peed in it, then presented it to my five-year-old sister. "For *me*?" Joyce said, her face aglow with childish trust, with innocent delight, while I felt instant regret.

She uncapped the tiny bottle and, sweetly smiling, smelled its contents. Then a hurt, knowing little frown crossed her face. "You *peed* in this, didn't you."

I laughed. But my sister got tears in her eyes, and I knew then, and forever, the cruelty of practical jokes.

I started school that fall. The first day my mother walked with me to the end of our long driveway and we stood waiting for the bus. Finally it came around the bend from the direction of the Huar place, not slowing as it approached. In fact, it didn't stop, the driver simply shaking his head at us as he drove by. "What's the matter with him?" my mother said. Then we walked back to the farm and my mother drove me to school.

She found out that the bus driver wouldn't stop at the head of our driveway because of the curve there and the danger from the poor visibility, but I could catch the bus at the Huars'. That meant an easy walk of about a quarter mile. It was a pleasant excursion along that quiet country road on fall mornings and again in the afternoons as the leaves turned and started dropping from the trees, and the birds—blackbirds, swallows—perched in long rows on the telephone and electric wires, flocking up for their flights south.

When winter started, my mother decided the walk to the bus-stop was too far if not dangerous in the sometimes sub-zero cold, so she arranged for me to ride with the milkman to school. That didn't last. The milkman was nice enough, attempting conversations with me as we drove from farm to farm picking up milk cans before we reached the school and he dropped me off. The trouble was he smelled strongly of alcohol, which made me afraid of him. (Never mind that my grandfather drank, and so had Oscar; a fear of drinking had been bred into me by my teetotaling parents and my grandmother.) Just reporting that the milkman had liquor on his breath, *first thing in the morning*, was enough for my mother. From then on, until warm weather came again, she drove me the two and a half miles to and from school.

I was attending the kind of one-room school still found then in rural America, though one by one they were being phased out. You walked in the door and found yourself in the cloakroom, where you left your jacket and hat and buckle overshoes. In the school itself, there was a blackboard and the teacher's desk at one end, then a potbellied stove, then the rows of desks, eight of them, one row for each grade. Out back were the outhouses, one for girls, one for boys. It wasn't pleasant having to go out there in the winter, but most of us had outdoor privies at home, so we were used to it.

The first day at school I was surrounded by the other boys in the schoolyard at recess. They were like a pack of dogs holding me at bay, teasing and poking at me. I closed my eyes and swung blindly at the circle of faces, slashing one of my tormentors with a fingernail. "Watch out!" he cried. "He scratches!"

They left me alone after that.

I waited for my mother after school. The teacher had to lock the classroom, but I could wait in the cloakroom, she said. My mother tended to be late. She might be in Hamel getting groceries or paying bills or visiting my grandmother. Or she might be at home and forget the time. One subzero day she was extremely late and my feet got cold and I started to cry. It was getting dark, and I paced the cloakroom, clenching my fingers inside my mittens and kicking the plank floor to ease the ache in my toes. I felt abandoned as only a child can. Finally my mother drove up, and then everything was all right.

Another time I was waiting for my mother, alone, I thought, when an older girl, perhaps an eighth grader assigned to clean the blackboard after school, came out into the cloakroom and said, "Do you want me to wait with you?" She was a dizzying female presence. Presently she said, "I have to go now," and I watched her cross the road in front of the school and run up a field road. After that I was sick with puppy love for her until school let out for the summer. I never saw her again after that, or didn't know her if I did. She grew up and got married, I suppose, and lived around Hamel for the rest of her life, for all I know.

Eventually my mother arranged for me to walk home with a boy whose parents' farm wasn't far from the school, and where I could play with the boy while waiting for her to pick me up. They were Dutch, a family of boys in which my friend was the youngest. His older brothers worked the farm while their fat, red-faced, pipe-smoking father, keeping company with his plump, hard-working wife, mostly just sat in the house and supervised, like some patriarchal Dutch farmer out of the tales of Washington Irving.

The teacher in that one-room school is faceless and ageless in my mind now, though I hear her strict voice and feel her discipline in front of the classroom. With eight rows of pupils and a daunting spread of them from pliable little six-year-olds to gawky, unruly thirteen- and fourteen-year-olds, you had to be strict, I guess, you had to be disciplined.

But in the late afternoons she used to read to us, a wonderful experience (and something I didn't get at home), and she taught *me* to read, literally poking the skill into my head, for which I thank her now.

Standing at the blackboard she taught us first graders the alphabet and about vowels and consonants and syllables. Gradually, we learned to recognize words. Then, one by one, we were called to her desk. She'd open a *Dick and Jane* reader or some other elemental text and point to the words and have us read them. "Sound them out!" she'd order, tapping the back of your head for emphasis.

That was her punishing method: having you lean over her desk, the reader before you, and sound the words out as she rested a hand on your head and tapped at it with a long-nailed finger. She tapped me in the same place, repeatedly with each mistake, until my head was sore and I cringed at each tap. I grew to dread my turn at her desk. Then one day, under her poking finger, as if by magic, the letters in the book formed words and the words formed sentences and I could read!

The teacher was reading to us to one warm, sleepy afternoon in late spring, at almost the end of that school year. The outside doors were open and there was the sound of a tractor in the field across the road. I couldn't see it at first because of the bank along the road, but then the farmer on his tractor appeared on the edge of the bank, directly in line with where I was sitting, and I saw his face, saw him wheel his tractor around and disappear again over the bank.

It was a scene so peaceful, so pastoral, and yet there was some menace in the air too, some vague unrest.

It was the spring of 1942, and America was at war.

I got sick that summer and wound up in a hospital. The probable cause was a diet of dill pickle sandwiches and Coca-Cola.

There was always a case of Coke on the back porch, brought home by my father, who was addicted to it, from his night job at the plant, and I took to helping myself to a bottle or two every day. My father would be sleeping and my mother didn't notice. I was a finicky kid, kind of sickly (from allergies, it turned out, like my sister Marcia), refusing most of the food on my plate. But I liked bread, my mother told me. *It was the one thing you'd always eat.* And I loved dill pickles. So I ate a lot of dill pickle sandwiches and drank a lot of Coke when my mother wasn't looking. Then something happened to my stomach. It got so I couldn't hold anything down, not even water. I'd throw up everything. Soon I was too weak to get out of bed, too sick to eat.

I was lying on a cot in the hallway outside the folks' bedroom one day when my father leaned over me and said that if I didn't eat I was going to die like Oscar. "Don't you wanna get better," he said, "so you can learn to drive the tractor?"

He'd bought our first tractor, a little Allis-Chalmers B ($570 with lights, starter and muffler, F.O.B. Milwaukee), and had promised to let me drive it when I was old enough. I was too sick to care now. I turned my back on him and went to sleep.

A day or so later I was lifted up by my father and carried out of the house. My head dangled below his arms as he walked to the car and laid me in the back seat for the drive to the doctor in Wayzata. *He's dehydrated, for Christ's sake. And it's your fault,* Dr. Devereaux told my mother.

I spent a week or so in Eitel Hospital in Minneapolis. They put me in the women's ward, on intravenous feeding, and I began to feel better. I liked being with all those women, most of them young and in confinement either before or after having a baby. I suppose I was comfortable with them because my grandmother and my mother's sisters always made so much of me.

I became the ward comedian. I could say almost anything and the women would laugh. The more they laughed, the "wittier" I became. They egged me on.

In the evenings, though, the ward was crowded with visitors and my folks didn't always come. I'd listen for their footsteps in the hall. When they didn't show up, I felt like crying. I could hardly breathe.

In the afternoons a nurse came in and drew the shades. Naptime. Bored senseless, I lay awake, waiting for the shades to be opened again, as the women around me slept. When I had the strength, I took to sneaking out of my bed to go to a window, slide under the shade, and watch life going on outside the hospital. People passed below me on the street. Beyond, I could see the pond and trees and grassy expanses of Loring Park. I was ready to go home.

One day a nurse said, "Tell me what you're hungry for. What's your favorite meal?"

"Roast beef sandwich with mashed potatoes and gravy," I told her.

It was brought to me. Delicious! I cleaned my plate. But a half hour later, I threw it up. So I was back on Jello and consomme, toast and poached eggs, for another couple of days.

Finally one evening the folks came with clean clothes to take me home. I jumped out of bed too quickly and almost keeled over. I was afraid I'd have to stay in the hospital. But then I was okay, and went around the ward with my folks to say goodbye to the women. I got hugs from them, and my folks heard how much fun I'd been, how I'd cheered them all up.

He's a card!

I was almost sorry to leave.

I remained sickly. The doctor said that having my tonsils and adenoids out might improve my health. So toward fall of that year, not long, I think, after our move to our second farm, I found myself back in Eitel Hospital, numb with fear but full of bright talk, "brave," the doctor told my folks, as I was etherised into a kind of nightmare echo chamber in which I heard Dr. Devereaux's booming voice and felt the pain of the knife, then woke in a hospital bed with a throat too sore to eat the ice cream my mother held out to me.

Back home, I talked funny for a few days, but my health improved.

2

The Mohrmann Place

THE MOHRMANN PLACE

Shortly after the move to our second farm, I went with my father to explore what he called Elder's Forty, detached acreage about a quarter mile away, rented from the man who'd sold us this farm. His name was Elder Mohrmann.

The front ten acres was field, the other thirty a mix of overgrown creek bottom, wooded knoll, and park-like little glades. Following my father into the hidden recesses of Elder's Forty was like venturing into a wilderness.

We had walked far, it seemed, into the wilds of Elder's Forty when, abruptly, we stepped out into open pasture. It extended across the line fence to a neighbor's meadow hayfield, likewise surrounded by trees, and the whole big opening, secluded there in the woods, was like a lost world my father and I had discovered. "Ain't this somethin'?" he said. A hawk appeared, soaring and screeching above us—we were near its nest, said my father. He raised his arms as if aiming a gun at the bird, said, "Bang," and the hawk veered away. It was hot and bright out in the clearing, and the grass was as tall as I was.

Elder's Forty was the first and probably the best of the wild places around the farm that would become my escape from it, my private Eden, an alternative world where I could live out my fantasies—where I could make up, as George Orwell did as a child, a story about myself.

The Mohrmann place wasn't far from our first farm, less than a mile as the crow flies, but at eighty acres, not counting Elder's Forty, it was twice the size of the Pepin place. And counting Elder's Forty and the grandfolks' forty acres outside Hamel, which my folks soon rented and eventually bought, we now had a total of one hundred and sixty acres of pasture and crop land. Later the folks rented additional land, ten acres here, twenty there, to raise more corn or alfalfa for our growing herd of dairy cattle or as a "money crop."

They paid some eight thousand dollars for the place, roughly double what they got for the Pepin farm. They borrowed the rest from the Hamel bank and from a rich, retired farmer in the area who lent money to his neighbors at an interest rate somewhat less or no more than a bank might charge. He'd give you a loan, if he liked you, after a bank had turned you down.

To fill the stanchions in the big barn the folks bought more cows—the grandfolks' herd of thirteen, in fact, since they were going out of farming—which increased our herd to twenty-six Holsteins and caused my father, soon after, to buy a third Surge milker.

It was the war that encouraged my father (my cautious mother had to be talked into it) to buy a larger farm and more cows. Milk prices were up—a national average of $1.73 a hundred pounds in 1942, compared to $1.22 in 1940—and farmers had been asked to double their production. Overnight, it seemed, the war had ended the Depression and improved the outlook for farmers—for everybody who wasn't off fighting. *It was a good time to make the jump.*

In the earliest black-and-white snapshots I have of the Mohrmann farm the buildings look recently painted and in good condition. (They were an off-yellow, the color of ripened grain or dry prairie.) The barn was a so-called bank barn, what you mostly saw on Upper Midwest dairy farms. Its ground floor was of fieldstone and built, as the name implies, into a hillside to form a walk-in basement. That was the cow barn, buried on three sides into a rise of ground facing southeast (the ideal situation), away from the prevailing weather, so that it was cool in summer, warm in winter.

The barn's expansive upper level was divided into floor-to-ceiling lofts or mows for storing hay on either side of the central "bay," which could be driven into through big sliding doors on the uphill side. A cistern, buried beside the barn on the up side, provided, like the system on our first farm, gravity-feed water to the cows' stanchion cups and the stock tank in the barnyard. The oldest pictures show the pipes angling down from the barn's eave troughs that originally fed rainwater into the cistern. But they were obsolete by then and were eventually removed because water could be pumped from the well above the barn into the cistern.

The farm's original well, a deep, rock-lined hole dug near the head of the driveway with pick and shovel, probably before the turn of the last century, was a trap we kids might have fallen into. Before it was filled in, Joyce and I used to lie on the rotting boards over the hole and drop stones through the cracks to hear them, after a long second or two, plunk into water. When the sun was right, you could actually see the dark water far below.

There was a small shed by the barn, later used as a chicken coop, and between the barn and the house a line of sheds that included a combination pump house and milk house; a big two-storey building that my father would convert into a granary and brooder house for chicks on the ground floor and a coop upstairs for the several hundred Leghorns that produced salable eggs and the folks' extra income; an attached toolshed; a machine shed; a corn crib.

Between the back of the barn and the uncapped cement silo, and attached to both, was a small building that included a basement feed room opening to the barn's manger and its two lines of stanchions, and a storage room above it where sacks of ground feed could be emptied through a chute into the bin below for rationing to the cows. There was a cart in the feed room for carrying silage to the cows after it had been forked down from the silo.

The sizable, two-storey house, built around 1890, had a bay window and two porches, a screened "back" one (leading into the kitchen; it was the entrance we mostly used), and an open, L-shaped, columned and gingerbread-trimmed veranda in front. It stood on a knoll above the other buildings and our short driveway, surrounded by shade trees and a stretch of lawn. The trees were box elders mostly, with a line of maples and big basswoods and one gnarled old willow out back. There was an outhouse in back and, off to the side and below it, an old smokehouse the folks never used except to store screens and storm windows.

Two brick chimneys stuck out of the house roof, one leading up from the cellar and the wood-and-coal furnace down there, the other, unused now, going up the north wall where the kitchen range once stood. It was the home of chimney swifts in the summer, and I used to watch them arrive twittering in the evenings to circle above the house until, one by one, they dropped into the chimney for the night.

The kitchen had an adjoining washroom and a pantry, and the kind of sink you saw in western movies or in the grandfolks' house in Hamel. The pump brought up water (and the occasional tree frog or stick bug) from the stone cistern in the cellar that collected rainwater from the roof by way of the eave troughs. It was primed with a dipper or two of drinking water out of the pail my father carried up to the house every day from the well.

Past the kitchen was the dining room with its rectangular table and matching chairs that perhaps had come with the house, and an ornate china closet with a counter and glass doors, built into an inner wall, below which was a linen closet that we kids could make a bed in, like one of those cabinet beds found in Dutch or Scandinavian houses. Off the dining room was the folks' bedroom, and beyond was the parlor with a sliding door that could close it off. There were doors out to the front veranda in both the living room and the parlor, doors to the outside in almost every corner of the house, in fact, which prompted my father to joke that the guy who built the house "musta been scared of his wife."

Upstairs there were two big bedrooms and three smaller ones that allowed separate rooms for my sisters and me, and for the hired man when we had one. There was a little door in the wall along the stairway that opened into a dusty storage space used mostly by us kids as a "cave" to play in, and a trapdoor in the hall ceiling that opened into a dusty, crawl-space attic with little diamond-shaped windows at either end from which hung shreds of ancient curtains. Clusters of bats, like dried fruit, hung from the joists, and there were rats, we discovered, in the dirt-floor cellar.

No electricity at first, but the folks would get it, as they had on the Pepin place, through the REA.

Before we moved in, they had the house fumigated for bedbugs.

❧

Saturday mornings, that first winter on the Mohrmann farm, my sisters and I left our beds in the cold upstairs to go downstairs and snuggle in with our mother. She was still in bed, waiting for the house to warm up from the fire our father had started in the furnace before going to the barn to do the milking. Our dad no longer worked nights at the Coca-Cola plant, though later he would "work out," days, during the winters. After our mother got up, we watched her comb out her long, lustrous black hair, which hung to the small of her back, then braid it into two thick plaits and coil them around her head.

One night my mother was awakened by the feel of something having jumped onto the bed. My father slept on beside her. "Puss, is that *you*?" she cooed, assuming it was our house cat, which sometimes slept at the foot of the folks' bed. She reached behind her and found the light switch. Turned it on. And stared into the beady eyes and quivering snout of a big rat.

"Oh!" she cried and, kicking at it from under the covers, catapulted the rat into the air. It landed on the wood floor with a plop, then scurried, with my mother in grim pursuit, out of the bedroom and across the living room and around the corner of the kitchen, then through the half-open stairway door into the cellar.

Somebody had forgotten to close the door after using the pot at the head of the cellar stairs.

I was on the pot late one night when something began rustling the scrap paper in the cardboard box beside me. I kicked at it and a rat jumped out, ran across my feet, and hopped down the stairs into the darkness. I finished my "job" pretty quickly then, and made sure to close the cellar door afterwards.

Then there was the Sunday, early that first spring, when we came home from church to find a skunk in our cellar. It had got in by way of the hatch-like doors above the wood bin. My father had cut firewood the day before and afterwards had left the doors open.

Carefully, we peered over the edge of the bin to see the skunk down there among the chunks of wood.

"Don't scare it," my father warned. "It might spray."

"What can we do about it, Mervin?" my mother asked.

"I'll think of somethin'."

The next morning I woke with a pounding headache. There was a thick, suffocating smell in the house. Then I remembered the skunk.

What had happened was that my father, digging carefully into the woodpile for pieces to start the fire that morning, uncovered the skunk. Up went its tail. He jumped back, avoiding its spray, but the cellar, and then the house, quickly filled with a scent so penetrating, so intense, that it seeped into your brain.

Joyce stayed home sick that day, but I went to school and sat among the titters of my fellow second graders in SS. Peter & Paul Parochial School in Loretto until Sister Corrine walked into the room, sniffed the air, and demanded, "What's that awful smell?"

More titters. I raised my hand.

"It's me, Sister."

"*Well*," she said. "I guess you're going home for a while."

At home the smell only got worse after the game warden was called (we thought he might have some neat, professional way of getting rid of the skunk) and he simply shot the animal and caused it to spray again.

"Hell, *I* could've done *that*," my father said. The warden left us, and my father carried the reeking body up the cellar stairs with a shovel and out of the house and buried it in the swamp pasture.

A day or two later a boxy old truck pulled into our yard, and a man in dirty overalls stepped out of it. The truck, and the man, gave off a powerful set of smells.

"I hear you had a skunk in your house," he said to my father. "What'd you do with it?"

"Buried it," Dad said.

"I collect furs. Mind if I dig it up?"

The smell of skunk quickly weakened in the house (or maybe we quickly got used to it), but lingered in the cellar for a year or more. All we could do about it was laugh.

One morning that first summer on the new place, my father woke me to get the cows.

"You're old enough now," he said. "But watch out for the bull. If he starts pawing and bellowing or starts toward you, duck under the fence and come and get me."

And so I began to learn the extent of our new farm, the lay of the land, by going to get the cows in the dewy early mornings and hot late afternoons. There was a small pasture just below the barn, and fields above it that could be opened for more grazing. Up along our east line was a wooded ravine, one of the farm's little wild places I would make my own, and beyond it a small field and then the much larger one that ran along our north border, past the woods there and our first farm. Peering out at it from those woods, and later working that field, going past the familiar buildings on our tractor, was like looking at my past life. I never knew the people who bought the Pepin farm from us, but I saw them driving away in their car once, the man and his grown son sitting up front, the woman in back, as I guess was the custom then among old-fashioned, male-dominated German or Dutch farm families.

Our permanent pasture was mostly a slough that ran along our west line fence and flooded in spring but dried up over the summer to provide "swamp" grass for our livestock. A cowpath led around the edge of it to a stand of trees and a grassy hillside where, in the afternoons, I usually found the cows lying in the shade and chewing their cuds or grazing the hillside. In the mornings they might be in the tall slough grass, and I sometimes had to wade through mud and water to get to them. I went barefoot, in rolled-up jeans or overalls, so wading into the wet was easy. The grass was alive with jumping frogs and the occasional, slithering garter snake. I was a little afraid of the snakes.

Beyond the trees the cowpath continued around past a brushy slope, then along the fence to a big old spreading willow in the midst of a rock pile. The rocks came from our field next to the pasture, rocks frost-heaved to near the surface every spring and discovered by the plow or spring tooth, to be gathered up and hauled by stoneboat to this pile by old Elder, and perhaps by his father before him. We would add to the pile ourselves over the years. You could find rock piles on every farm, along the edges of fields, sometimes in the middle of them, where they appeared as brushy islands and revealed, on inspection, a gully or an outcrop or a "low spot," piled with field stone and overgrown with wild vegetation.

☙

Our nearest neighbors, the Jensens, were a pair of strapping Scandinavians—he Danish, she Swedish, I think—without children. But just over the nearby Cemetery Hill lived the Hermans, who had kids our age, we discovered. Their farm was next to my aunt Betty and uncle Gerry's place, and the cemetery, on the knoll beside the road, held the graves of the area's first white settlers, including Mr. Herman's parents and grandparents, their names spelled Heermann as in the old country. Mrs. Herman kept the grass mowed up there and tended the plots. It was a nice overlook from which to view the surrounding country.

There were two boys and a girl in the family. The older boy, Nick, was a big blond kid a year older than I and a year ahead of me in school. His sister Laura was Joyce's age, and their baby brother Johnny was Marcia's age. They were German and Protestant—Lutheran, I think—and enrolled in the one-room public school a couple of miles down the road, while we, Catholic and considering ourselves French, would go to the sister school in Loretto that fall. We were "different," we thought. Never mind that our father was a somewhat reluctant Catholic (forced to convert to marry our mother), was half German, and we had a German name. What's more, Loretto was a German community, as Hamel was French. *Oh, don't talk like those old Germans,* my mother would tell me when I began to say "dis" and "dat" for this and that.

Soon after our move to the Mohrmann place, my father and my uncle Gerry started farming together. They formed a partnership, K & L Farms ("K" for Klatte, "L" for Leuer), and had the name printed on their cheques.

Your dad and I just tried to help each other out. Neither of us could find a hired man because of the war. If you were single and able bodied, you were in the service.

My father was the city kid, still learning how to farm. My uncle had grown up on his farm and inherited it, had most of the machinery needed to work their combined acreage (Uncle Gerry had one hundred and seventy acres to our home eighty), and made a point of reading up on modern farming methods. He'd even, after high school, attended the "farm school" at the University of Minnesota.

It was decided my uncle would do most of the field work and my father would do the milking. This resulted, however, in some resentment. *Your uncle Gerry had his Sundays off, for one thing, and I never did.*

Eventually there was hard feeling—on my dad's side, anyhow. He was a proud man who up to now had learned what he knew of farming from my grandfather and from watching the neighbors, noticing when they planted, when they harvested, how they put up hay; learned from his mistakes mostly, learned things the hard way. Now he was being told how to do things by my uncle, a man six years younger than he was. Never mind that the man had been raised on a farm, had the know-how, and was better educated and maybe even a little smarter than my father.

They had a showdown one day, and K and L Farms broke up. Of course, as family, my dad and my uncle continued to associate, but after their quarrel there was always a certain tension between them. Once, years after they'd farmed together, they had a shouting match in which they dredged up all the "shit," as they called it, from their K & L days and threw it at each other.

Years after that, they were driving somewhere in my father's pickup when he was moved to say, "How about it, Gerry? The two of us ain't getting any younger. What say we let bygones be bygones?"

The two old men turned to each other and shook hands.

Your uncle got tears in his eyes, for chrissake!

It was close to Christmas, our first or maybe our second Christmas on the Mohrmann place. I was seven or eight years old and upstairs, alone, shivering from the cold up there and from my awful daring, standing in front of the unlocked but forbidden door of the "junk" room.

It was one of the two large bedrooms upstairs and served as our storeroom. Ordinarily it wasn't closed to us kids. But near Christmas that year we were told to stay out of it.

"Why?" we wondered.

"Just stay out of there," our mother said.

It was maybe a week before Christmas and we had our tree up in the back parlor of the house, a big spruce or pine as high as the ceiling, bought in the lot behind Fortin's Hardware in Hamel and carried into the house, screwed into its stand, and decorated by my mother and us kids. I don't think there were presents under it yet, but soon there would be and they'd be wrapped presents, with our names on them, left for my sisters and me to heft and shake and try to guess what was in them. We'd get to open them on Christmas Eve—the night *before* Christmas—because Christmas morning there'd be the milking to do and then Mass to attend in Loretto, or more probably in Hamel, because after church we'd go to the grandfolks' for dinner. At the grandfolks' we'd get more presents—at least one from them, then maybe another from your godfather or godmother, and finally one from the aunt or uncle who had drawn your name.

Most exciting, though, were the presents we'd get from Santa Claus. And they'd arrive, unwrapped, on Christmas Eve too, before bedtime, as if the considerate old elf knew we had chores to do and church to go to on Christmas morning.

So there I was, in front of the junk room, unable to resist my curiosity and yet hoping against hope I wouldn't find anything. Shivering, I opened the door, went to the closet, and found a pile of presents, none of them wrapped. There was a wooden doll's house whose roof could be lifted off so as to expose its three or four little rooms in which there was miniature furniture; there were a couple of cutout books of paper dolls. Obviously these were meant for my sisters. For me, obviously, there was a wonderful toy battleship, two or three feet long and made of wood and metal and painted navy gray. It was a scale model of the *Oklahoma*, on which my uncle Billy had served before it was sunk during the Japanese attack on Pearl Harbor. It had spring-loaded guns on its foredeck that shot wooden bullets and a little seaplane that could be spring-catapulted from an upper deck.

My heart was pounding. I put the toys where—and exactly as—I'd found them and tried to forget them, tried to will what they suggested out of my mind.

I had my little chores to do that Christmas Eve: calves to feed, probably, and bed down for the night. The older calves were in their pen, and I suppose I threw hay to them and spread fresh straw over the soiled straw. The newborn calves were tied to a manger along one wall of the barn, and they were fed by putting them beside their stanchioned mothers to suck. I wasn't old enough to do that yet, wrestle a calf to its mother. But I was learning to wean the calves, getting them to drink milk out of a pail by dipping my hand into the milk and then letting the calf suck my fingers while drawing or pushing its head down to the pail. They soon got the idea.

My uncle Gerry was there (we were K & L Farms by then), his herd together with ours in the barn (he kept pigs at his place). He must have thrown hay down to the cows and bedded them down with straw while my father finished the milking and washed the utensils in the milk house. Then the men lifted the half-dozen or so ten-gallon milk cans onto the stoneboat and hauled them with our Allis-Chalmers up the hill from the barn to the cooling tank in the milk house. Afterwards Uncle Gerry went home to his own Christmas.

I got to the house to find my smiling mother and impatient sisters waiting. "Hurry up!" Joyce said. "*Santa's come.*"

"Wait for your dad," our mother told us.

When our dad came into the house, we all moved to the parlor, whose sliding door had been closed to encourage Santa to pay his visit. One of the folks slid it open.

There, under the tree, was the toy battleship. There was the doll's house and other toys I'd seen already, mine and my sisters'. All unwrapped, all, I had to realize now, from the folks. There was no Santa Claus.

My folks sat smiling together on the couch, and my sisters were jumping with surprise and joy around the tree, clutching their presents from "Santa," while I had to pretend that *I* was surprised and full of joy. I had to pretend I still believed in Santa Claus, though my heart was breaking. I was sick with disappointment, sick from my loss of innocence.

I like to think it wasn't only that I had to hide my disappointment or the folks would guess I'd "peeked," but that, unless I acted my part in what I knew now to be a lie, I'd destroy my sisters' happiness and, what's more, *the folks'*. I saw my parents' beaming faces, saw the light in their eyes that reflected the light in our eyes, though *my* eyes nearly filled with tears, and maybe I knew then that our happiness was also theirs.

Sister school

For Joyce, starting first grade at SS. Peter & Paul parochial school in Loretto was her introduction to school. For me, entering second grade and coming from a one-room public school, "sister school" was like starting over.

The public school had been like play school compared to SS. Peter & Paul, which at first was a little like Navy boot camp would be for me thirteen years later. Things were regimented at SS. Peter & Paul. The nuns were as strict as boot company commanders.

They were School Sisters of Notre Dame, an order that had originated in Germany and whose regional mother house was a hundred miles south of us in Mankato. They were good, demanding teachers, ex-farm girls or from small Midwestern towns, who were not unkind (some even had a sense of humor), but who ruled, quite literally and in the time-honored way, with a ruler. *Whack* on the palm of your outstretched hand when you were bad. Otherwise, they would sometimes threaten to send you to Father Kaufman and his rubber hose. Nobody, so far as I know, was ever disciplined by the parish's old priest, and I suspect his fearful rubber hose was an invention like the bogeyman my grandmother said would get me if I wasn't good.

What I dreaded more than Father Kaufman was the school cafeteria. It was staffed by women of the parish—volunteers who took turns—and the parish being German, we were fed a lot of wieners and sauerkraut, which wasn't bad, but also such strange concoctions as "scalped" potatoes, strongly laced with onions. It was good peasant fare, I suppose, as was one of my favorite dishes at home, an unthickened stew of beef, potatoes and carrots—rutabagas too, when my grandma made it; and, yes, a bit of onion and garlic—that we called *bouillon*. You ate it on a plate rather than in a bowl after mashing together some meat and vegetables from the pot with your fork, then pouring the stew's "gravy" over the mess and seasoning it with salt and pepper. Delicious.

I was nauseated, though, by much of what they served in the school cafeteria, and so devised a way of getting past the nun whose job, along with maintaining order during the meal, was to inspect our trays and ensure that we'd cleaned our plates. What I'd do was pick at my food until Sister wasn't looking, then squash it under my plate.

I got caught at it one day and was made to stay in the cafeteria until I'd eaten my food, while the other kids were allowed to walk, not run, outside for noon recess. I sat alone and stubborn in the cafeteria until the school day ended. I heard the rumbling overhead when classes were dismissed. Then heard the rustle of her stiff habit as Sister Bernice, arms tucked firmly into her gaping sleeves, came down the stairs and glared at me.

"Okay, Mister, you got your way *this* time. Take your tray to the kitchen and go."

Eventually I realized that this witchlike, controlling figure, the Sister Superior, took an interest in me. I used to catch her looking at me sometimes with a quizzical glint in her steely eyes, and toward the end of my schooling at SS. Peter & Paul, she said to my mother, "I wonder what he's going to be?" She might have wondered because by that time I'd shown I would rather read than anything; by the time I reached the fourth grade I'd ask to borrow *National Geographics* from the library to take home on weekends (*Do you have homework? No, Sister. You sure? Yes. All right, then*). By the time I reached the seventh grade and was in her classroom, I'd ask, when my work was done for the day, to go to the back of the room for a volume of the *American Encyclopedia* to page through. By the time I reached the eighth grade, I'd gone through the set. There were so many interesting facts about the world, its people and places—other people and other places, other periods in history, elsewhere than where I was. I was wondering myself what I was going to be.

Though we all feared her, we laughed at Sister Bernice behind her back. Somebody noticed she picked her nose and ate her boogers, and after that we watched for it, saw her dig into a nostril behind her handkerchief and then, after a furtive glance at us, pop what was stuck to her finger into her mouth. We'd exchange looks then, and mean little childish grins.

The school day began with separate musters in the three classrooms. Then we were marched in a body to the church for eight o'clock Mass. That first day the boy walking beside me introduced himself. When we reached the church, he said, "Let's sit together." His name was Dan, and we were instant friends.

Back in school, classes started with a prayer and, with the war going on, the Pledge of Allegiance to the Flag. After the noon recess there was more prayer. In the afternoon, following regular studies, there was catechism. At 3:30 we cleared our desks and spilled out of the classrooms to fetch our hats and coats from the wall hooks in the central hall. Then we filed past the nun at the school entrance, Sister Superior or her second-in-command, who made sure we were carrying books home to study—no library books allowed, unless you were caught up with your work. Since our school books had to be covered with oilcloth and marked appropriately—English, Arithmetic, etc.—I soon learned to sneak out library books under bogus covers.

Outside the school we were lined up, two-by-two again, then set to marching like little soldiers down the sidewalk toward "uptown." We didn't have uniforms, strictly speaking, but the girls were made to wear dark skirts and white blouses and we boys had to dress in slacks and ties. *No overalls or blue jeans.*

Looking back you saw the nun's black figure in the middle of the street between the school and the convent house, arms folded in her sleeves, the church steeple rising like a phallic symbol behind her, making sure we held ranks. Once we reached the main street, we could scatter. Whenever we grew lax in our ranks a nun, playing drill sergeant, marched beside us to within a hundred yards or so of the main street. I don't think any of the nuns were allowed to venture uptown except Sister Superior when she got their mail at the post office.

It was at sister school, more so than at the little one-room school I'd first attended, that I noticed the different smells the kids had. Each had the smell of his or her household, I guessed (apart from the barn smell of farm kids, which anyway hardly registered), their family smell, the smell that came from what they ate at home and how often, or seldom, they bathed. Some actually stank (usually those with "green" teeth from never brushing); other kids' smells were only interesting, even intriguing, while the smell of the girl who became my grade school heartthrob was positively alluring. What was *our* smell, I wondered?

We traded rides to school with the Schmidts, whose farm was directly across the road from Aunt Betty and Uncle Gerry's place. For a week my mother drove us to school in the mornings and the Schmidts picked us up in the afternoons. Then they switched. It saved on rationed gas.

The Schmidts were childless but fostered orphan kids, usually a girl and a couple of boys old enough to help with the farm work. There was a regular turnover of those kids throughout my grade school years. They were different from us, damaged, my mother said, hurt, "immature," sometimes "retarded." I never felt quite comfortable around them.

There was a boy named Marvin who drooled and would do anything on a dare. Once, riding home with the Schmidts in their boxy, 1928 Chevrolet, which they always ran well under the reduced wartime speed limit, only allowing an increase in speed when they coasted downhill in neutral, Marvin said, "I bet I could jump out and run alongside the car."

"Go ahead, " I said, eager to see what would happen.

What happened was that Marvin opened the back door, stepped out, and was jerked comically out of sight. Darlene Schmidt, sitting up front beside her husband, Joseph, who was driving, turned around at the sound of the back door opening, then yelled, "Stop the car!" Joe applied the brakes. Behind us Marvin was picking himself up from the gravel, pants torn at the knees, his hands bloody. Still he was laughing, drooling and laughing. I was afraid he'd tell on me, but he never did.

Later there was a girl, Josephine, short and dark, tomboyish but pretty, whose body, I began to notice, was developing in disturbing ways. One summer evening we got into a wrestling match on the Schmidts' lawn. We were rolling on the grass like young animals, without a clue as to what was happening, except that I liked the soft feel of her breasts when I brushed against them (they were surprisingly large), the swell of her behind, the smell of her sweat. We were panting, as much from excitement as our exertions, when Josie suddenly stopped and drew away from me. She gave me a funny look. How old were we? Twelve, maybe. Old enough.

Ivan was another of the Schmidts' orphans, whom the Schmidts got, I think, after the war. He was a year older than me, bigger and stronger, handsome and athletic, and he tried to be my friend. I envied him—he was a good ball player, for one thing, while I was a lousy one—and I should have been flattered by his attention. I regret now my standoffishness and wish I had been a true friend to him because I felt his need for it, his orphan's loneliness, his sadness, his vulnerability. But I was afraid of him. I was afraid of his "fits."

He must have come to the school as a seventh grader, and so was in Sister Bernice's home room a year ahead of me. So it wasn't until the next year, when I was a seventh grader, that I witnessed Ivan's epileptic seizures in a classroom.

Before that, I'd seen them on the playground. The first time one happened, I saw a group of boys gathering at the far end of our side of the playground and thought a fight had started. Then I saw girls running from their side to join the group, and a boy passed me yelling, "Come on!" and I followed. Inside the circle we made, there was Ivan, writhing on the ground, foaming at the mouth, his face purple and contorted. We all just stood there, watching something we'd never seen before. Ivan was gone from us somehow, gone from himself, it seemed. "Step aside," Sister Bernice said when she came briskly up, forced her handkerchief in Ivan's mouth, and dispersed us.

Ivan's *grand mal* seizures on the open playground were disturbing enough, but in the closeness of the classroom they gave you a headache. Once one of his fits started, you realized you'd been waiting for it all along, wanting it to happen, finally, so as to get it over with.

You'd hear his hoarse cry behind you. Then the thump of his big body as he fell against the floor. You'd turn to see his rigid, foaming-at-the-mouth convulsions between the desks, so powerful he once lifted a row of the screwed-down desks right up off the floor. The violent part of the seizure might be over in a minute or two, but then he lay unconscious, his breathing raspy and terrible. We'd shy away from him and stand staring from the front of the room. "Don't be afraid," Sister Bernice told us, but you could see his fits upset her as much as they did us.

She always put something in Ivan's mouth, her handkerchief again or a ruler, to keep him from swallowing his tongue, she said. After he quieted, she got a couple of the eighth-grade boys to carry him into the hall. They covered him with our coats. Then one of the boys was assigned to stay with him until he came to. The hallway door was closed and we went back to our work

After a while the hall door opened and Ivan entered the classroom looking dazed and sheepish, almost ashamed. Nobody ever knew what to say to him.

We were still trading rides with the Schmidts, so I got to know him better. I would have avoided him otherwise. Once he took me into the shed where the Schmidts kept their car and pointed up to the ceiling. There, under the peaked roof, its runners laid across the open joists, was the Schmidts' old cutter from the days when they hitched a horse to it to get around in the winter.

"Watch this," Ivan said. He went to a corner of the shed and climbed the studs and blocking of the uninsulated and unpanelled wall to the several boards laid across the joists. Then he walked to the sleigh and sat in it.

"Come on up," he called. I did, and sat on the cracked leather seat beside him. He looked around and grinned. We were hidden up there with the dust and the cobwebs. "Nice up here, ain't it?" he said, then "Shhh!" Crouched down in the seat, quiet as the spiders in the corners, we heard the shed door open and Darlene Schmidt call out, "Ivan, you in here?" Ivan put a finger to his lips and shook his head. Darlene closed the door and went away. Ivan grinned again, and I knew he'd shown me one of his secret places.

I was pretty well adjusted to sister school by then. Gone were the winter mornings of my first years at SS. Peter & Paul when the cold and school seemed to go together and I lay in bed after my mother called me, working myself up to being sick.

"Why aren't you up?" my mother would call up the stairs.

"I'm sick!"

"Are you *sure?*"

"Yaas!"

"All right then. I guess you'll have to miss school."

I don't think Joyce ever tried that ploy. When *she* stayed home, she seemed really sick.

I enjoyed those cosy days in the house with my mother, not bothering to get dressed but staying in my bathrobe and slippers, perched in front of the heat register in the dining room, soaking up the acrid warmth from the coal-burning furnace in the cellar and listening with my mother to radio soap operas: *Ma Perkins, Stella Dallas, Our Gal Sunday*. I'd get to eat lunch with my mother, something good like Campbell's soup and grilled cheese or peanut butter sandwiches.

Pretending or willing yourself to be sick wasn't the only way to miss school. The weather—in winter—gave us a treat sometimes. There were snowstorms in the winter—a couple of bonafide blizzards, if we were lucky, the snow blowing across the road, making drifts like waves in the fields, piling against the snow fences—that had us listening to the radio in the mornings for the school closings. When SS. Peter & Paul was named, Joyce and I cheered. There were nights I went hopefully to bed with the start of a blizzard outside. Snug in my blankets I'd listen to the wind moaning around the eaves. I'd hear the rattle of the driven snow against my window and feel the drafts, the cold breath of the storm seeping into the house through its uninsulated walls. And I'd fall asleep with the pleasant thought that there might not be school the next day. With luck, we might even be snowed-in.

❧

I became an altar boy at SS. Peter & Paul after my First Communion, when I was ten. Sister Bernice instructed us new boys. She took us into the sacristy, which smelled exotically of wine and incense, opened a closet to show us our vestments, showed us the priest's vestments, and ran us through our movements at the altar during Mass. Our "homework" was to memorize our Latin responses.

Even today I can recite most of them. When the priest began the Ordinary of the Mass by intoning, "*In nomine Patris, et Filii, et Spiritus Sancti. Amen. Introibo ad altare Dei*" ("In the name of the Father, and of the Son, and of the Holy Ghost. Amen. I will go to the altar of God"), the server replied, "*Ad Deum qui laetificat juventutem meam.*" ("To God, the joy of my youth.")

"*Judica me, Deus, et discerne causam meam de gente non sancta: ab homine iniquo et doloso erue me,*" the priest continued. ("Give judgment for me, O God, and decide my cause against an unholy people, from unjust and deceitful men deliver me.")

And the server's reply: "*Quia tu es, Deus, foritudo mea: quare me repulisti, et quare tristis incedo, dum affligit me inimicus?*" ("For Thou, O God, art my strength: why has Thou forsaken me, and why do I go about in sadness, while the enemy afflicts me?")

And so on.

In the first half of the Mass, the Mass of the Catechumens, "The joy of my youth" refers to the joy of those who will be born again in Baptism; the "sadness" has to do with the recognition of one's sins.

The Holy Ghost is now called the Holy Spirit, a more direct translation from the Latin, but "ghost" then was perfect: it suggested all the mystery and terror of religion.

Until I was about twelve, I used to wander the woods around our farm, inspired by the teachings at SS. Peter & Paul and such movies as *The Song of Bernadette* and *Joan of Arc*, half yearning for, half fearful of, heavenly voices or a vision of the Virgin Mary. And during Mass I couldn't watch the priest's elevation of the Host at the Consecration, after the thin white disc of unleavened bread had been miraculously changed through transubstantiation into the Body of Christ— *Take ye all and eat of this: For this is My Body!*—without the awful possibility of seeing Christ Himself. I wondered how the Eucharist tasted until I'd experienced my First Communion (like cardboard, actually), and wondered too how the wine tasted (only the priest drank from the chalice, on behalf of the congregation) after it had become the Blood of Christ.

The first year or two I attended SS. Peter & Paul, stuffed with stories from the nuns about hell and satanic possession, I'd lie rigid under my blankets at night, waiting for the bed to levitate or to hear the sound of the devil's rocks pelting the house. I was afraid to fall asleep because I might die in the night, and saying another Act of Contrition didn't help much:

O my God! I am heartily sorry for having offended Thee, and I detest all my sins, because I dread the loss of heaven and the pains of hell . . .

I said my prayers every night, kneeling beside my bed: an Our Father, a Hail Mary, and an Act of Contrition. Then I'd ask for God's blessing on my family and myself: "God Bless Daddy and Mama and my sisters and brothers and aunts and uncles and Grandma and Grandpa and all my cousins and make me a good boy."

I recited that each night, without self-consciousness, until I left home at age twenty and began to lose my faith, to "fall away," as my mother and our parish priest feared I would.

I was touched, though, by the liturgy. I loved the organ music, the singing of the choir at High Mass. And I loved the beauty of the English translation of the Mass, which I recited from the pews along with others in the congregation when I wasn't serving at the altar with the priest. SS. Peter & Paul, unlike St. Anne's in Hamel, practiced the "dialogue" or "community" form of the Mass in which the Ordinary, composed of the Mass of the Catechumens or Learners (converts preparing for Baptism in the early church) and the Mass of the Faithful (in which only baptized Christians in the early church were allowed to participate), was recited in English by the congregation while it was being conducted in Latin at the altar. In this way the Mass wasn't simply "heard" but was participated in by the congregation as it had been during the first ten centuries of the Church.

Our "tool" for this practice was a useful little book called *My Sunday Missal*, by the Rt. Rev. Msgr. Joseph F. Stedman, director of the Confraternity of the Precious Blood in Brooklyn, N. Y.

I especially loved the English translation of the Last Gospel in Father Stedman's missal, the Gospel According to St. John with its rhythms, its incantatory repetitions, which ended the Mass, and which began: *In the beginning was the Word, and the Word was with God; and the Word was God. He was in the beginning with God. All things were made through him, and without him was made nothing that has been made. In him was life, and the life was the light of men. And the light shines in the darkness; and the darkness grasped it not.*

In Latin the words were gobbledygook. In English they raised the hair on my head.

When serving Mass you went early to the church sacristy and got into an ankle-length black robe called a cassock and a white blouse called a surplice. Then you helped the priest get into his vestments: the shawl-like amice, the white linen robe called the alb, the cord-like tasseled cincture around his waist. Then the maniple, a long band of cloth worn over the left arm; the stole, a long scarf worn over the neck and crossed at the chest; the chasuble, a sleeveless, knee-length cape covering front and back with a hole in the middle to pass the head through. All these vestments, according to our *Missal* originated as items of clothing in the ancient Near East. Alb is the Latin word for "white," and was an ordinary outer garment worn against the hot sun in the Near East. The maniple was a cloth worn on the arm to wipe away dust and sweat. The chasuble, from the Latin *casula*, "little house," was worn as a protection from the weather and surely is the ancestor of the poncho. The outer vestments are in six liturgical colors, and worn according to this or that feast or church season. White is the symbol of light, joy and purity; red of blood and fire; green of hope; violet or purple of penance; rose of joy during penitential seasons; black of mourning.

You served alone for Low Mass and with another boy for High Mass. There was Low Mass every weekday, both High and Low Masses on Sunday, and High ("midnight") Mass on Christmas Eve. Low Mass was relatively quick, just the basics, but lengthened by the sermon or homily on Sundays; High Mass was longer, seemingly more reverent, with chanting of parts of the Mass by the priest and singing responses by the choir. The choir also sang hymns during the Mass, many of them beautiful, some only tedious.

Special feast days—Pentecost, Easter—called for processions, outside in good weather, inside in bad, led by the priest and his servers. Then there were weddings and funerals to serve at on occasion, and "Stations" (The Way of the Cross) every Friday night during Lent.

During summer vacation, each boy was assigned a week of service at the daily Mass. For me that meant walking (until I had a bike) the three and a half miles to Loretto in time for the eight a.m. service. If I got to the church on time, Father Michael sent me to the vestibule, where its rope hung down from the steeple, to ring the bell. That gave me a bit of fun. To start the bell ringing, I'd jump up to grab the rope and hang on as the bell's rocking weight lifted me up and down like Quasimodo in *The Hunchback of Notre Dame*.

When I was late, as I often was after first having to get the cows for the morning milking, I got one of

Father Michael's "looks." What was it? Disgust? Disappointment? It made me feel worse than any bawling out I might get from my dad.

Father Michael, who'd replaced old Father Kaufman, was a small man with a James Cagney strut. He took charge of the playground—the "boys' side," that is, while the girls kept to their side where a nun's black figure stood watch during recess. He pushed us out of our bunched-up inactivity during recess like a livestock shipper driving cattle up the loading chute with an electric prod, and soon we were all striving to meet his sarcastic, perfectionist demands. Depending on the season, he organized us into teams to play baseball, touch football, hockey. What's more, he played *with* us, always taking the weaker side. He had the little man's determination to stand as tall, somehow, as other men. He'd drink in the taverns after church on Sunday, which scandalized the more righteous in his flock, who eventually got him transferred. Once, so the story goes, in one of the taverns, he was insulted by one of his drunken parishioners and offered to fight him. They might actually have traded blows, but for another parishioner, a German farmer with arms like sides of beef, who took the man outside and taught him to respect his priest.

In the school yard Father Michael fixed on me, the smallest boy in my class, as a kid who needed encouragement. He'd toss me the ball, but I'd usually flub it. He was the first of those men who seemed more interested in me than my own father, who became mentors, father-substitutes, men whom I feared and admired as

I did my real father and strove not to disappoint but whom, more often than not, I disappointed.

Father Michael drilled us in Latin, insisting that it be spoken quickly, succinctly, correctly. Being quick was his style: he could say Low Mass in exactly forty-five minutes, timing himself with the watch he placed on the altar. He was correct, it seemed, in everything, and at the same time recklessly impulsive. One spring day during noon recess he got into his car, said "Jump on!" to me and a couple of other boys, and headed out of Loretto toward Highway 55 as if racing to a fire. We stood on the car's running boards, our arms hooked through the open windows, whooping into the wind and hanging on for dear life.

We were taught to carry the priest's big missal from the Epistle to the Gospel side of the altar during a High Mass as if it were the colors for some flag-raising ceremony on a military post, and how to hold it—just so—for the priest to read or sing the Latin. For the Offertory, when we carried the cruets of water and wine to the side of the altar (water added to wine symbolized the human and divine in Christ, our redemption by the blood and water that had poured from Jesus' spear wound on the Cross), Father had us empty the wine into his chalice, then add a drop of water, a drop only, so as just to conform to the liturgy. (Under Father Kaufman, the ratio had been the opposite.) The wine had a peculiar, pleasant smell; it wasn't ordinary red or white wine, I discovered later, but sherry, with twice the alcohol content of wine.

One Sunday after Mass Father Michael asked my friend Dan and me to go with him to see Nazareth Hall, a Catholic boys' school in St. Paul that was a kind of pre-seminary. Dan and I were looking forward to the next fall when we'd enter the public high school in Mound; never mind that it was a seat of godlessness and temptation. It was obvious to Dan and me that Father meant to save us from that, to entice us into the priesthood. Going to Nazareth Hall would be a waste of a good Sunday, but neither of us could say no. Under our mothers' smiling approval, we climbed into Father's car and let him drive us the thirty miles or so to what later a former student with masturbatory inclinations and Joycean wit called "the semenary."

We nosed about Nazareth Hall all that deadly dull afternoon, watched some loud, swearing boys play pool in the dormitory rec room (it was as if they were trying with their profanity to convince us, and themselves, that they were just like other boys), and chatted with the Fathers. But from the moment I stepped inside the walls of that place, I saw myself imprisoned there. Those walls, though I didn't quite realize it then, were a manifestation of how I was beginning to feel about the farm.

Wartime

No bombs fell on us. No tanks rolled across our fields and into our villages. Neither the "Japs" nor the "Krauts" invaded us. But we were in the Second World War and it hung over us, the news of it continually on the radio and in newsreels at the movies, in magazines and newspapers. There was the organized austerity of the home front (the rationing of all luxuries and quite a few necessities); the feel of a gigantic, nationalistic effort underway; the overwhelming dedication and patriotism and anxiety. The dark excitement.

And also undeniable evidence of wartime prosperity: the Depression ended, full employment—and for dairy farmers, better milk prices, bigger milk cheques.

In March 1943 my mother went into Eitel Hospital in Minneapolis and came back, some days later, with twin boys, whom she named Michael Albert and Mark Hamel, their middle names a play on her father's name. Mike and Mark weren't identical twins, but people outside our immediate family couldn't tell them apart.

One evening, later that spring, my father put me on our little Allis-Chalmers tractor and set me to "dragging" a recently plowed and disked field in front of the barn as he did the milking. At eight years old and small for my age, I couldn't reach the clutch pedal to put the tractor in or out of gear, but I could steer, he said, and he had me climb up on the seat with him and sit between his knees. He put the tractor into gear, adjusted the throttle, and we started down the middle of the field, pulling the drag harrow, a raft-like set of steel teeth. He turned at the far end of the field and drove along the fence, being careful not to hook any of the posts with the harrow, until we were back where we'd started. He took us around again, running along the outside of the center swath, then along the inside of the swath by the fence. Then he stopped the tractor and said, "You see how to do it? Just go back and forth, keeping to the outside over here and the inside over there. I'll keep my eye on you."

He started the tractor moving again, said, "Grab the wheel," then stood up from the seat behind me and let himself down between the rotating rear wheels and the following harrow. He hung on to my seat for a moment and kept pace with the tractor, then adroitly stepped clear. I turned to see him walking toward the barn.

Perched up on the tractor and struggling to steer it, I'd see him come out of the barn at intervals and stand for a moment watching me. That was a comfort. Meanwhile I began to see how my staggered back-and-forth loops, inside and outside the drag's swaths, were closing the gaps between the swaths, breaking up the clods of dirt and smoothing the soil as if with a giant rake. Getting it ready for planting.

I'd nearly done the field and was getting close to the fence on the outside loop when my father walked over from the barn and met me, swung up behind me on the tractor, and took the wheel. He ran as close to the fence as you could get without catching on it, and so finished the job for me. "You did good," he said. His words filled me to bursting.

My mother had "complications" after the twins were born and went back into the hospital. When she came home again, she had a hired girl.

Her name was Marge and she was two months shy of thirteen, big and strong, with sandy blond hair, a pleasant smiling face, and a big laugh. She talked—and laughed—a lot.

I got a crush on her immediately. She looked, I would decide, a little like Ingrid Bergman in *For Whom the Bell Tolls,* a movie I saw about this time with my folks at the Wayzata Theater.

Aunt Betty and Uncle Gerry, after suggesting my father hire a girl to help my mother, had found her on a neighboring farm, where Marge's mother herself worked as a hired "girl" after leaving Marge's father and coming down from the "sticks" outside Little Fork,

Minnesota, near International Falls and the Canadian border, with her three daughters and a son. Marge was the second oldest of the brood.

She came to us at the end of June, and might have gone back to her mother in the fall, except that her mother was unsure of where she'd be working that winter and wanted Marge to have the stability of our place long enough to finish grade school. So she was still with us, fortunately, when my father had his crippling accident.

It happened on Wednesday, September 8, 1943. My father fell into the silo filler, a noisy, infernal machine with a furious maw of whirling knives and a roaring fan that chopped stalks of corn into bits and blew them up a pipe into the silo. He had most of one foot chewed off and a damaged leg, but he might have been more horribly injured or even killed had my uncle Gerry not been there to save him.

They'd been filling our silo using our Allis-Chalmers and two wagons, with my uncle hauling the corn in from the field and my father unloading.

Earlier, a field of immature corn had been cut with the binder. It was probably my uncle's corn binder (we would get our own eventually), a one-row machine pulled behind a tractor and run off its power takeoff that cut the corn stalks, tied them with sisal twine into bundles, and dropped them in bunches onto the field. Before this a three- or four-row swath had been cut by hand through the middle of the field over several days and the stalks fed to our cows after evening chores. The swath allowed you to start binding without driving over any corn.

The silo filler was Uncle Gerry's machine, and it ran off the pulley drive of his old Case tractor. Our silo was a round cement tower like a medieval battlement, smaller than some of our neighbors' and without a cap, some ten feet across and about thirty feet high. When the silo was filled, its capacity could be increased by adding a length or two of snow fence along the top. The chopped corn quickly settled and fermented into a juicy, fibrous, smelly mess, which the cows loved as my grandfather loved whiskey.

My father was on top of a load, lifting up bundles of corn, cutting the twine, then throwing the stalks into the machine when suddenly part of the load gave way and he fell onto the conveyor.

Fortunately, he fell feet first. And fortunately my uncle was still in the yard, about to pull out with the empty wagon, when he heard my father yell and saw him fall. In an instant, Uncle Gerry was off the tractor and literally diving for the control bar to reverse the conveyor. When he lifted my father away from the machine, they both saw that his left foot had been cut off to the instep ("adding a little meat to the vegetables," as my father joked later), and that there was a nasty set of puncture wounds along his shin into which bits of dirty overall and long underwear had been driven by the steel teeth that pulled the corn into the chopper.

The wound hardly bled. That was the funny thing about it.

That was how Uncle Gerry remembered it. He drove my father—and my half-hysterical mother—the twelve miles or so over dirt and blacktopped roads to the doctor in Wayzata. All the way, my father sat white and silent in the back of the car, holding tight to the towel wrapped around his mangled foot. At Doctor Devereaux's he got a shot of morphine and his wound was cauterized. Then he was driven to Eitel Hospital to be operated on.

His shredded foot was trimmed and sewn up, and it soon healed. Gangrene, though, developed in his punctured leg. The doctors tried penicillin, one of the new "wonder" drugs, but there was talk of having to amputate. When Joyce and I went with our mother to visit him in the hospital, we looked at each other and crinkled our noses. Our father's room stank.

Uncle Gerry did our chores. Other people, aunts and uncles and the grandfolks, pitched in. There was a benefit dance in Hamel that helped pay the doctor bills.

My father was out of the hospital, *doing the milking on crutches,* my mother recalled, when he posed with us for the pictures taken October 17, 1943, on our front lawn with the milk house and a corner of the barn in the background. It was a Sunday after church, and my aunt Clarice was there—she's in one of the pictures. Aunt Clarice, the second youngest of my mother's sisters and very 1940s classy looking, was still living at home in Hamel, awaiting the return of her fiancé (later my uncle Gene Miller), then in the Army somewhere in the Pacific.

She took the group picture of us, my father and mother, us kids and Marge. Ma's holding one of the twins, Marge the other. There's a snapshot of my father alone, smiling on his crutches. He might have been a soldier in civvies, invalided out of the war. Instead he was just a farmer, wounded in the hazardous business of farming.

Marge stayed with us through that winter and the following summer. (That must have been the summer I watched her and my father one evening from my upstairs bedroom window; she was helping him load or unload something in the yard; they looked like a couple; I was sick with jealousy.) Then she went to stay with an aunt and uncle in Iowa and started high school there. She came back to us for her sophomore and junior years at Mound (she'd turned Catholic by then, inspired by my mother and following a private meeting with her in the folks' bedroom from which she and my mother emerged, tearful and smiling), then went to a Catholic school in Sleepy Eye, Minnesota, for her senior year.

We paid her, she's said, ten dollars a month, room and board, and only a dollar a week during the school year. *I thought that was a lot of money.*

She was more than just our hired girl; she was my mother's friend and companion and in fact she became something like an adopted member of our family. In part, she and my mother got along so well together because they were both card-playing "demons."

Saturdays we used to hurry up with the housework so we could play cards through the afternoon. Could she ever work! She was the best hired girl I ever had.

The summer after her high school graduation, she returned from Sleepy Eye to work for Aunt Betty and Uncle Gerry, and that fall entered the convent of the School Sisters of Notre Dame—but not before sowing some wild oats. She tried smoking (*I doubt I smoked a whole pack that summer*), got drunk once at a wedding, but these "tests" of her religious vocation only confirmed it. I was in the eighth grade and in the home room at SS. Peter & Paul when Marge (now called Sister Marge and a Candidate in her order) spent a day or two in our room as a classroom observer before practice-teaching fourth, fifth and sixth graders in the intermediate room. She was smiling and friendly, still the Marge we knew inside her long black dress and white-trimmed black veil, but she wasn't our hired girl anymore. When she was received as a Novice, she became Sister Corona; later, under Pope John XXIII's enlightened reforms, she was given her baptismal name back and became Sister Marge again.

For a while after his accident my father tried wearing a heavy brace, composed of metal shafts and hinges, leather straps and a wooden "toe," that extended from his mutilated foot to above his knee. He'd gotten the thing by way of his accident insurance. But it was tiring to wear, he found, and actually worsened his limp. Eventually he threw it in a closet, where I dug it out sometimes and strapped it on.

Instead of wearing the brace, he took to filling the toe of his left shoe with rolled-up socks and just walked on the remaining heel of his bum foot for the rest of his life. In the morning, he hardly limped. By afternoon, though, near exhaustion, he limped markedly, almost as if he had a peg leg, sometimes pushing himself so furiously that, years later, observing him at work on the golf course he had created from what had been part of our farm, my university professor brother-in-law—jokingly, but behind his back—called him "Ahab."

The stump of his bad foot, a flap of skin stitched over severed bone, was like a huge callous that he massaged at night to relieve the ache. He could tell when it was going to rain because his foot ached sharply then, and asleep in bed he was often startled awake by a cramp in that leg.

Being in pain much of the time didn't improve his temper, and he and my mother fought at times, mostly over their money and how he wanted to spend it. He might shout at her, then stomp out of the house, jump in the car, and drive crazily down the road in a cyclone of dust. "He'll be back," Ma said. Sure enough, fifteen or twenty minutes later, our car would appear on the road, being driven so slowly as to raise no dust at all. It would pull into the yard and stop, and our father would climb out of it with a sheepish grin on his face.

A couple of years after losing a foot, my father nearly lost his other leg in an encounter with our Holstein bull. The bull, too big and belligerent to let loose in a pasture anymore, was kept in his steel pen in the barn and only turned out into the fenced yard for exercise. We'd raised him from a calf, and my father thought he could handle him until one day, when Dad was inside the pen cleaning it, the animal's bellows and snorts and angry pawings convinced him to get out of there. He backed toward the pen's horizontal bars and was rolling through when the bull charged, pinning my father's good right leg against the bars with his massive, dehorned head.

Powered by adrenalin, my father raised his manure fork and stabbed it into the bull's neck. *He didn't budge. I had to drive the fork in again, halfway up the tines, before the fucker backed off and I could roll free.*

He crawled up to the house and startled my mother. Again, there was a frantic drive to Dr. Devereaux in Wayzata, and again gangrene set in. And yet again, penicillin saved him.

Uncle Gerry had a wolfish-looking mongrel shepherd named, what else, Shep. I petted him when nobody else did at my aunt and uncle's, and he started following me home. Finally he just stayed at our place and Uncle Gerry said I could keep him. "But watch him. He likes to run cattle." So I was never to take Shep with me when I got the cows.

That first summer we had Marge, she came with me one evening to get the cows. I was thrilled to have her along, to show her how I did it. We were starting up the fenced lane from the barnyard when Marge looked back to see Shep behind us.

"Nuts," I said. "I forgot to tie him."

Shep came sidling up to us and Marge patted his head. "Aw, let him stay. I'll watch him."

The cows were on Elder's Forty. To reach it you followed the lane to our east boundary and the crossroad. Then you walked down the crossroad to another lane, lined prettily with saplings and a rail fence, that ran along the front field of Elder's Forty to the wooden gate at the head of the back pasture. The gate opened onto a dusty or muddy clearing, depending on the weather, where the cows usually lay, chewing their cuds, waiting to be herded home. If they weren't by the gate, they might be grazing on the creek bottom, where there were grassy clearings in the woods. Or they might be at the far end of the pasture, in the big clearing, grabbing last mouthfuls of the knee-high grass as I approached.

Wherever we found them that day, we got them started toward home and I ran ahead at the two places where the cows might stray—where the pasture lane came out onto the crossroad, and again where the cows crossed the county road and entered the lane going down to the barnyard.

Shep walked obediently behind us. We paid less attention to him, I guess, as the cows plodded along, heading with growing excitement toward the barn, where they knew feed was waiting and they'd be relieved of their milk.

That was the trouble. Nearing the barn, the lead cow, bag swaying, broke into a trot. That caused the other cows to begin trotting, and of course Shep's chase instinct was triggered. He leaped past us and started nipping at the cows' heels. "Shep!" Marge and I yelled. "No, Shep. NO!" But he didn't stop. The herd stampeded. Through the dust they kicked up we saw some animals break through the barbed wire along the lane and others run on and through the fence around the barnyard. It was terrible to watch, and when at last Marge and I caught hold of Shep and tied him up and had the cows rounded up in the yard, we could count the casualties.

Some of the cows had only scratches along their flanks, but there were others whose heavy bags and swollen teats were torn and bloody from the barbed wire.

Marge went right up to the house and told my mother. "It was all my fault," she said.

My father wasn't home but would be pretty soon. "He'll blame Ross," my mother told Marge. Then to me she said, "Get on your bike and ride to your aunt Betty's. Stay there till I call to say it's all right to come home."

I was full of guilt and fear, crying in choking, hiccupping gasps as I pedalled my bike to my aunt Betty's. She was cheerful and comforting and fed me supper. Uncle Gerry, when he came in, said we'd have to shoot the dog. Presently the phone rang, and Aunt Betty answered it. "Your ma says it's safe to go home now," she said with a crooked little smile.

My father wasn't in the house when I got there. He was in the barn, tending to his nervous cows, disinfecting their wounds, their shredded teats and udders, then milking those that would let him, torn teats notwithstanding. Surely, being upset, they were holding their milk, which would cause a loss in production, a reduced milk cheque.

Over the next days my father had to watch the cows' teats and udders, look for the swelling to go down, look for signs of stringy, lumpy or bloody milk; watch for mastitis, in short, which would mean a ruined cow. After the war penicillin, in self-lubricating sticks for insertion into the teat canal, became available for such conditions as mastitis. But there was nothing like that now.

I waited for my father's reaction to what I'd let happen to the cows, as terrified of it as when I'd broken the windmill, back on the Pepin farm, but he never even bawled me out. What kind of trial lawyer's eloquence had my mother brought to bear on my father? What persuasive pleas of guilt had Marge offered? Shep wasn't even shot. He died a year or two later, more or less naturally, after being run over while chasing a car.

Late one sub-zero night toward the end of the war strangers knocked at our door and woke us up. The sky was red outside, reflecting on the snow, and the yard was lit by a huge fire between the house and the barn. It was our two-storey chicken coop, with our four hundred Leghorns inside.

The strangers were a couple of young guys driving home from a dance who saw the fire on the horizon, drove toward it, and found our place. It was one or two in the morning, and so cold the snow squeaked under the wheels of vehicles that began to turn into our yard, and under our feet when we stepped out of the house to watch the fire; squeaked, that is, except close to the fire where the snow was melting.

We kids were able to stand outside in our slippers and pajamas, in the heat of the flames, and watch Hamel's volunteer firemen fight the fire. About all they could do was try to contain it and save the buildings on either side of the coop, the machine shed and combined milk house and pump house. As the fire subsided, it got too cold for pajamas and slippers and we went back into the house and watched through the kitchen window from on top of our mother's sewing machine. The three of us, Marcia, Joyce and I, crowded the top of the sewing machine and saw the fire die down, leaving the chicken house a caved-in, smoldering ruin. The men were rolling up the hoses, under the yard light, when we went back to bed.

The next morning, before school, I walked through the ashes and wisps of smoke, among the charred boards and the singed and partially cooked carcasses of all those chickens, all our good laying hens, that had constituted the folks' egg business. There would be no more egg route for my father on Saturdays, when sometimes I accompanied him and the storekeepers gave me penny candy. After the fire, my father worked out in the winter. We still kept chickens, but only fifty or so for our own use, in the small hen house by the barn.

Sometime after the fire, and perhaps because of it, my mother gave birth to a stillborn. *It was a little girl. We buried her in a coffin the size of a shoe box.*

Just about everything that you couldn't grow locally or make at home was rationed during the war —sugar, coffee, cigarettes, nylon stockings, gas and oil, tires. The speed limit was reduced to thirty miles per hour in the daytime, twenty-five at night. There was a Goodrich tire slogan that went:

> *Sure, slow driving*
> *is a chore.*
> *But saving tires*
> *helps the war!*

There were "junk drives." On the Pepin farm I had helped my father collect old tires from the dump on the place. The money we got was mine, he said, and went into the Hamel bank for me, but pretty soon had to be "borrowed." Any kind of scrap metal, especially iron and steel, was needed for the war effort. Scrap rubber, rags, discarded rope, old burlap bags, waste cooking fats, waste paper and tin cans—all was saved and recycled. It was an environmentally-healthy practice that would be forgotten as soon as the war was over.

Joyce and I were sent into the fields a couple of times to gather milkweed pods for their fluff inside; it was used in life preservers, we were told. Our mother flattened her empty tin cans and threw them into a corner of our back porch to be taken, periodically, to the depot in Hamel; there was a clatter every time she threw another can onto the pile. My sisters and I watched her clip coupons from her ration books before going into Hamel or Minneapolis.

Despite wartime rationing, despite the fact that the folks were in debt—then and through all their years of farming they lived mostly on credit—I never had a sense that we were poor. We weren't, in fact: we owned our land, our livestock, our machinery—that is, so long as the folks kept up their payments at the bank—and always had plenty to eat from the bin of potatoes in the cellar, the raw milk from our cows, the eggs and meat from our chickens, tender veal from our butchered calves—the little bulls mostly, except one raised occasionally to replace a bull getting too old and heavy to breed the cows without driving them to their knees. When a cow stopped producing, Dad butchered it and had the tough beef ground into hamburger, or he shipped the animal to the stockyards in South St. Paul.

At the start of every winter the folks brought home a crate of oranges, which we kids ate, two and three at a time, as if starved for vitamin C. They brought home sacks of apples, some from the grandfolks' couple of trees. For sweets we had Ma's cookies or cake, canned fruit, and sometimes pie and ice cream, the pie's delicious crust made with lard, the pure rich ice cream ordered from the creamery by way of the milkman. For a quick, between-meals snack, we made a mush of graham crackers and milk. That satisfied your sweet tooth.

In winter, for cheap entertainment, the folks joined a card club, and went off once a week after evening chores to play cards at one of the neighbors'. Club members took turns hosting the card parties. When it was the folks' turn, it was a special night for Joyce and me. We were allowed to stay up as the house filled with farm couples that came in out of the cold after they'd done their own chores. They piled their coats on the folks' bed and sat around our kitchen and dining room tables and an additional card table or two as the cards were shuffled for games of cribbage or gin rummy or penny ante poker. Joyce and I circulated among the tables, listened to the boisterous talk, and waited for the end of the evening when cake and ice cream were served, with coffee for the adults and milk for us kids. The milk was from that "saved" in a gallon jar every evening during the milking by my mother—or by Marge, when we had her—and carried up from the barn to our refrigerator, which the folks still called the "icebox." Finally, the couples hunched into their coats and swept back out into the cold to their cars and Joyce and I were shooed upstairs to our beds.

We had visitors, mostly relatives, who came out to our farm on weekends or in the evenings, especially in summer; but also, occasionally, in the evenings after chores, a neighboring couple with kids stopped by, which gave my sisters and me somebody to play with. A favorite game, after dark, was called Ghost. It was a form of Tag, only much more exciting because you got a scare out of it.

We'd draw straws or flip a coin to determine who'd be the first "ghost." Then the rest of us would press our faces against a tree in the yard and count to fifty while the "ghost" ran behind the house and hid. Then, bunched together, we walked around the house chanting:

Star light, star bright,
I hope I don't see the ghost tonight.

We kept that up until the "ghost" appeared, jumping at us from out of the shadows. The girls screamed and all of us ran wildly about as the ghost tried to tag one of us to be "it."

It was a fun game but had its hazards. They were our mother's wire clotheslines, strung between the trees in the yard. One night the oldest Huar girl, tall for her age and playing the ghost, leaped out of the dark at us, ran into a clothesline, and practically knocked her front teeth out.

My mother went occasionally into Minneapolis to shop. I used to ask to go along, and suffered her seemingly endless strolls up and down Hennepin and Nicollet Avenues to enter this or that store, notably Dayton's or Donaldson's, to accumulate bags of purchases, which I helped carry, for the treat afterwards of Chinese food at the Nankin or the buffet lunch at the Forum. Then, best of all, we attended a matinee movie. With my mother I must have seen most of the Hollywood musicals of the 1940s, including all of Esther Williams's underwater ballets, which my mother especially liked—I suppose because she herself couldn't swim. Musicals were a favorite of my mother's, but she also liked mysteries, thrillers, historical romances. She liked movies, in short, and so did I; we were the movie buffs in the family.

Maybe that's why she took me into town with her, and not Joyce (Joyce is absent in my memory of those trips into Minneapolis, though probably we took turns going with our mother). Invariably we'd miss the start of the movie, so at the end we'd wait for the next show and watch it until we reached the part where we'd come in. Then, if the show had been especially good, my mother would lean over to me and whisper, "Should we stay and see it over?"

Those were the years when my mother sat down sometimes at the old upright piano in a corner of the dining room and played "Clair de Lune" or "Rustle of Spring" for us. The piano, which had been hers when my mother was a girl and took music lessons, had been hauled out from the grandfolks' house in Hamel. At my mother's urging, I took piano lessons for a while from one of the nuns at school and learned to play, by ear (I never learned to read notes), some of the songs in Thompson's first-grade book, including one of the simplest, "Swans on the Lake." I can still play it.

During the war milk prices continued to rise, and in our area they rose even higher than the national average. According to the U. S. Department of Agriculture, the average price to farmers for a hundred pounds of milk, which had been $1.73 in 1942, jumped to $2.15 in 1943, and to $2.30 in 1944. By then, the *Farm Journal* was already warning of a postwar fall in prices.

My father remembered getting $1.50 to $2.50 per hundred pounds of Grade A milk from the Loretto creamery before the war. During the war, he said, the price rose to $3.50, and one incredible month the folks got $4.50 a hundred. Hauling cost twenty-five cents a hundred and was deducted from your milk cheque. The milkman, though, who collected your full cans in the morning and returned the empties in the afternoon, couldn't always get through. There were drifted roads in the winter, muddy, rutted roads in the spring, when his truck couldn't reach you. Then you got the milk out yourself, by horse and sled or stoneboat through the snow in the winter, by tractor and wagon through the mud in the spring. Always, the struggle was to keep your bacteria count down, stay on Grade A.

After the war, it got harder. The price of milk did go down, while regulations increased. The milk inspector came around when you least expected him, and he could lower your grade or shut you off altogether if he found your manure pile too big or an unclean barn or dirty utensils in the milk house. There were funny stories about the response of some farmers to the milk inspector. One, said to be a little crazy, locked the inspector in the barn; another, considered sane, after following the inspector around as he totted up a growing list of infractions, lost all patience when the man bent over the cooling tank to smell the water. The farmer raised a foot and shoved the inspector into the water.

One day the Schmidts picked us up in Loretto after school and told us excitedly there'd been a plane crash in Will Jensen's field by the road. We'd see it when we drove by.

We saw the smoke first. Then we rounded the corner at the top of Arens's Hill and there, in the Jensens' field—not far from their house—was the smoldering wreckage of a B-17. A Flying Fortress! A big piece of the fuselage was crunched against a sidehill. Other pieces lay scattered over the field. You could see where the skidding plane had gouged a furrow through the hay stubble. There were Army trucks lining the side road leading to the Jensen farm, and a crowd of uniformed men around the site.

The crew, in training out of Wold-Chamberlain Field near Fort Snelling, had experienced engine trouble and bailed out of the plane, allowing it to crash. They'd floated down over the countryside to land miles apart from one another.

For a couple of weeks we could drive by that field after school and see the wreckage still there, and a soldier with a rifle on his shoulder pacing to and fro in front of a tent. After the wreck had been hauled away, piece by piece, and the soldier and his tent were gone, the path he had worn into the field, along with the scar made by the crashing plane, remained. They were there until Will Jensen, later that fall, plowed all evidence of the crash under.

The war held back the mechanization of farming that had started after the Civil War, had reached a peak during the Depression, then was suspended by World War II rationing. If you still had horses, as we did, you kept them. If you had a tractor, you used it as little as possible, though farmers were rationed more gas than their city cousins. Still, for a while longer, farming wasn't all that different from my grandparents' day.

Threshing, for instance (pronounced "thrashing"). The first time I heard a threshing machine approaching our farm, it was out of sight and I didn't know what it was. What I heard was a mechanical *clap-clapping* to the east of us, over the near hill. Then I saw it, a big old steam tractor and a huge, galvanized metal machine being pulled behind it, the two like joined behemoths heaving ponderously over the hill, *clap-clapping* toward our place at about five miles per hour. The rig belonged to a farmer who made a circuit around the area each harvest season, stopping at the farms on his list. The neighbors got together with their wagons and teams to thresh as once they'd gathered to pick one another's corn or shock one another's grain or raise a barn.

The tractor and thresher pulled into our yard and the thresher was parked and levelled where my father wanted his straw pile. A belt was attached from the power pulley on the steam tractor to the drive pulley on the thresher, with a twist to keep it on the wheels, and the tractor was positioned to take up the slack. Before this my dad and uncle and some of the neighbors had been out in the grain field—we grew mainly oats, not much wheat; the big wheat fields were on the prairies west of us—to break up the shocks and turn the bundles' damp undersides to the sun to dry. Now the wagons and teams were in the field, loading up with bundles, and pretty soon they started coming in and lining up on either side of the machine's feeder apron. Then the tractor pulley was engaged and the thresher started. The old farmer poured a sticky fluid on the inside of the belt to help bind it to the pulleys, and the men on the loads started forking the bundles into the machine's maw, heads first, so the grain was correctly shaken from the stems and sifted down into a bin and the stems and chaff—the straw—blown up and out a big pipe with an adjustable mouth, like the neck and head of a mechanical dragon, to form a pile that would be as big as our barn by the end of the day. In the field, meanwhile, men called spike pitchers loaded the wagons.

The day was filled with noise, with itchy dust and chaff, the clapping tractor with its spinning flywheel, the shaking thresher, the blowing straw, the grain dropping from the bucket that measured and counted the bushels into the bin that funnelled down to fill a truck or wagon box or a gunny sack held by a man. There was a man on a little platform beside the strawpipe, cranking the wheels that swung or lifted it, angling the mouth to shape the pile; and there were men in trucks or on horse-drawn wagons hauling bundles in from the fields, and the grain truck or wagon hauling loaded sacks or loose oats to the granary, everybody working, each knowing his job. It was something to watch, a ritual of cooperative labor that harkened back to the oldest days of farming and was near its end, though nobody knew it yet.

My mother would be in the house making dinner. The kitchen table was pushed into the dining room to join with the expanded dining room table, which had all its leaves in place. Joyce and I helped lay the tablecloths and carried food to the table. Outside, we helped set up the wash table, with a milk can of water brought up by one of the men from the pump house and a dipper beside it, and a basin with soap at one end of the table and towels at the other. When dinner was ready, I ran to tell my father, who raised and lowered his arm as the signal to shut down.

Now was when the men found out what kind of a meal the woman on the place had prepared for them. Feeding a threshing crew was an unspoken form of competition among farm women. They secretly vied with one another to pile the men's plates with fried chicken or roast beef or baked ham or pork chops, fresh green vegetables, corn on the cob, Jello salads, canned this or that. Then treated them to desserts of cake or pie, strawberry shortcake, ice cream. Word got around as to who put on the best feed. I think my mother did okay in this department.

After the heavy meal, the men stretched out under the trees in the yard for a half-hour nap. Then they went back to work. The idea was to finish by chore time, but the evening milking was often late on threshing day. After the war, we had a succession of hired men, but never one my father trusted to do the milking. He trusted *me*, after *I* became his hired man, by which time we had our own combine and he'd send me home to do the chores while he kept combining. I was beginning to hate the work, but I liked the fact that he knew I could do it.

May 7, 1945. V-E Day. The folks were having the house wallpapered, proof, no doubt, of their wartime prosperity. A couple, the man small and skinny, the woman big and fat, had been hired to do the work. We kids spent most of the day in the house, watching the skill that went into wallpapering, listening to my mother and the couple talk, listening to radio reports about the victory in Europe.

Then V-J Day. News of it reached us late on Wednesday, August 15, 1945. That was "free show" night in Hamel. It was the night when old movies were shown, outdoors, against the side of one of the taverns—the treat of the week for us kids, and for adults too, through those austere summers during the war.

Come Wednesday, we prayed for good weather. After evening chores, if it was clear, we drove into Hamel and laid out blankets in the parking lot of whichever beer joint was sponsoring the show that week. We sat there on the ground, swatting mosquitoes, or walked around, visiting with neighbors, until the guys from Minneapolis arrived with a 16mm projector and reels of film. As soon as it was dark enough, the projector started up and a full show, just like in the theaters, began: newsreel, cartoon, a short feature (some wry little piece by Robert Benchley, say, or a James A. Kirkpatrick travelogue, badly photographed in color), and then the main feature. Mostly they were early talkies, with a muffled or crackling soundtrack. There were old Laurel and Hardy comedies, newer Abbot and Costello ones, The Marx Brothers, The Three Stooges; 1930s romantic comedies, "B" movies of various kinds.

It could turn chilly on those summer nights, and we kids wrapped ourselves in blankets and pulled them over our heads while continuing to stare at the magical images on the screen, which was just a bed sheet nailed to the outside wall of the tavern. Behind us was the whirr of the projector, and in the projector's beam of light we saw swarms of insects, flitting bats.

There was a long intermission between the short subjects and the feature, which gave the hosting tavern a last chance to make some money—gave the adults time for another drink or two at the bar and us kids a chance to buy pop and other treats. But then the show started again, whirring on until midnight or later—until, despite ourselves, we were overcome with sleepiness and longed to be home and in bed.

Hamel was especially crowded that August night in 1945. There was muted hilarity in the air, talk of the news that a terrible new weapon, the atomic bomb, had been dropped on a couple of Japanese cities, and that the "Japs" were about to surrender. We were all waiting. I was waiting for the show to start; the adults, I guess, were waiting for the expected announcement from President Truman.

It got dark, and still the men with the projector and the night's movies hadn't come out from town. People were standing around in the parking lot beside

Pepin's, milling out into the street, talking and laughing. It was so late by now, they were just shapes under the street lights.

Then, from toward Minneapolis, we heard approaching car horns. They drew closer, honking continuously, sounding like a wedding party. Suddenly the town was filled with the honking cars, their drivers shouting out the windows.

"THE WAR'S OVER! THE WAR'S OVER!"

In the delirium that followed I kept hoping that people would settle down finally and there'd be the usual free show. But then, as the bunches of excited talkers gradually thinned, and as car after car started up and drove out of town, I was forced to realize, with sinking disappointment, that there would *be* no show that night—that everybody except me, apparently, had forgotten about it.

Postwar

The two snapshots are dated July 1946 and show a couple of gigantic loads of loose hay on our old Model A truck. They were taken on separate days, judging from my father's change of clothes, probably during second-crop alfalfa that first summer after the war.

In one picture my father stands alone in front of the truck at the head of our driveway. The truck's sun-discolored windshield is cranked open and its hood panels have been removed, indications of hot weather. The load rises in a peaked stack behind him. He's in old dress pants and a short-sleeved shirt, his feet spread, tanned arms loose at his sides, his tanned face a little weary looking under an old fedora that looks vaguely like a tropical pith helmet. His left leg rests slightly ahead of him and his weight is thrown back on his right foot. Ever since his accident three years earlier, he's favored the remainder of his left foot.

The other picture, taken from some distance away (perhaps by me), has both my folks beside the truck, their figures dwarfed by another huge load of hay they've brought in. My mother's white and overweight, my father lean and dark. She drives the truck for him these days. During the war and gas rationing, we'd used a wagon and horses to make hay, and I'd learned to drive the horses while my father loaded. But at eleven years old now I'm still too small to reach the truck pedals so my mother drives, straddling the windrows, shaded in the truck cab but suffering the heat of the engine wafting back at her through the open windshield. Meanwhile my father, out in the hot sun and standing in the truck bed between its high wooden racks, meets the hay spilling off the loader being pulled behind the truck, and spreads it with a fork. He spreads the hay until it piles up over the top of the racks and the reach of the upright loader. Then piles on more.

The two pictures remind me of the one and only time my father laid a hand on me.

We still had our horses, a team of big Clydesdales that together were more powerful than our little Allis-Chalmers. They could lift the hayfork, larger than the one we'd had on the Pepin farm, its four, clawlike tines spread out and set into the hay, when the tractor couldn't. I drove the horses. And I knew to say "whoa" when the rope the horses were attached to slackened, which meant the fork had snapped into the catch at the top of the barn and was moving along the track over the mow. I'd hear the *whump* after my father pulled the release rope and the hay dropped into the mow.

I'd been riding with my mother in the truck that day, to be on hand for the unloading, but then it was time for my programs, *Hop Harrigan, Jack Armstrong, The Lone Ranger*, those oldtime radio serials (fifteen minutes each episode) that were aimed at every boy in America. They came on between four p.m. and six, Monday to Friday, and I hated to miss them.

"Oh, all right," my father said. "You can listen to your programs. *But watch for us. Don't make me have to come and get you.*"

The folks drove off for a last load before chore time. I stood behind our floor-model radio set into the bay window of the house, listening to my programs and watching for our truck to turn into the yard. Eventually I tired of this, and lay with my head on a pillow at the foot of the radio, my usual listening position, assuming I'd hear the truck when it drove in. As it happened, I didn't.

What I did hear above the exciting drama on the radio was the sound of the screen door being flung open and the thump of my father's bum foot striking the kitchen floor.

Before I could get to my feet, he came through the door from the kitchen, his face contorted, and was on me. Gripping my collar and the seat of my pants, he jerked me off the floor and threw me, in a horizontal spin, across the dining room. I landed in a heap, picked myself up, and attempted to walk with some dignity out of the house. I heard his thumping behind me and hunched my shoulders. When I reached the porch, he kicked me down the steps. I stood up again and heard my mother yelling as she approached the house, "Mervin! That's enough! You're going to hurt him!"

I didn't care. Walking stiffly to the barn, I felt my father behind me and waited for a push or another kick, but he was done. He hitched up the horses for me, handed me the reins, and said "*Drive.*" I did that, walking beside the horses' scrabbling hooves, my head only inches from the groaning, quivering rope taking the strain of lifting hay off the truck that, if it broke, might snap my head off. *Go ahead and break,* I thought. *He'll be sorry!*

I was on the floor in front of the radio again, stubbornly listening to the rest of my programs, when my father came into the house for supper. He stuck his head into the dining room with what can only be called a shit-eating grin and rolled a fifty-cent piece across the floor to me.

"There. Go buy yourself somethin'."

His lame apology.

Accepted.

We were still without indoor plumbing—wouldn't get it until Joyce and I were in high school. Washing, bathing and elimination, therefore, were done in the age-old way.

Monday was washday, when my mother rolled out her wringer washer and sorted the dirty clothes into light and dark piles. She was lucky in that she had a machine with an electrically powered agitator and wringer; there were neighbor women still using the old crank-operated washers. But she used a stick, as they did, a cut-off section of an old mop handle, to fish the clothes out of the rinse water before she fed them, piece by piece, through the wringer. The stick, from years of swirling it around in the wash water, was as smoothly worn as a piece of driftwood.

There was no drain in the room off the kitchen that was my mother's washroom, nor a source of water in there either, so the business of washing and then rinsing the clothes involved carrying pails of hot water from the kitchen to fill the washer (water pumped from the cistern at the kitchen sink and heated on the electric stove), then draining the used water into the same pails and dumping them into the sink. My mother, though, had help after the twins were born—from Marge, then later from another hired girl, and eventually from my sisters.

Drying the clothes was another job. In summer my mother hung them outside on wires stretched between the trees on our lawn (the same wires we kids had to beware of when playing Ghost after dark in the summer), and in winter she hung the clothes inside the house from clotheslines stretched along the upstairs hallway. It was so cold up there in winter, the clothes took forever to dry.

Then the ironing. My mother spent most of her evenings at it, the board set up in the dining room where she could listen to the radio as she stood sprinkling with water and then flattening and smoothing with her iron virtually everything she'd washed, including the handkerchiefs. She was usually ironing when I came up from the barn after chores, and I liked to sit by her then, to read or listen to the radio with her: *Suspense*, *Fibber McGee and Molly*, whatever was on. With my nose in a book and often only half-listening to the comedies she especially liked, I took comfort in the rhythmic stamp of my mother's iron.

Saturday was bath night. Water was heated on the stove and then poured into a washtub for people to share, each in turn, so that the last to use it, though tepid water had been subtracted and more hot water added, bathed in what looked like old dish water.

The tub was placed in the kitchen by the heat register or in the relative privacy of the pantry. Coming up late from the barn, I often got third dibs on the tub, after Joyce and Marcia. The folks bathed in a fresh tub, my mother first, usually, and then my father. The twins were bathed by my mother in the kitchen sink.

In fact, until we got too big and heavy for her, our mother bathed all of us kids that way. She'd stand you in a basin of warm water in the sink, soap and rinse you, including washing your hair, then carry you to the kitchen table to towel you dry. The last time she did this for me, I must have been eight or nine. She just managed to lift me to the table, then said, "You're getting too old for this—in more ways than one."

Eliminations were something else. The outhouse, which neither my mother nor my sisters liked because of the stink and, in summer, because of the wasps and spiders in there—in winter it was just too cold for their tender fannies, my father said—was mostly used by my father and me, and the hired man. The women mostly used the chamber pots kept under the beds for "number one," and we all, on occasion, used the pail-size pot at the top of the cellar stairs for "number two." I took to pissing through the screen in my bedroom window, like other farm boys I knew, writing my initials with the stream so that the letters, *R K,* became rusted into it.

An improvement over the chamber pot was something called a sanitary toilet. In effect, it was an adult-sized potty chair warmly placed behind the furnace in the cellar (where, sitting on it, you might hear a rat scuttling in a corner), but the five-gallon pail under the seat was always full, it seemed, and had to be lifted out and carried up the cellar stairs and through the house without slopping, and dumped in the outhouse. Nobody ever wanted that job. But somebody—me, when I got old enough—had to do it.

Of course for my father and his hired man, there was always the barn. That explained the roll of toilet paper kept on the ledge above the calf manger.

Eventually we got indoor plumbing. But my mother, after nagging my father about it, had to get things started herself.

Mervin said he was going to put in the water and sewer lines so we'd have a bathroom. But he never did it. He said he'd have to start by digging the trench for the drain pipes, but I knew darn well he'd never get to it. He'd never have the time, for one thing. And so one day I looked in the paper and found some guy with a back-hoe that I called up and he come and dug the trench and Mervin comes home and there it's done!

Meanwhile, my mother had put an ad in the paper to sell her piano, and before my father got around to putting a plank across the ditch to get in the house, a man came by after dark that evening to look at the piano.

He knocks at the door and Mervin opens it and says, Oh-oh, I hope you didn't fall in our ditch! I did, the man says.

Then one evening Aunt Stella and Uncle Rex came out from Minneapolis with Grandma Loomis for a visit. After dark, with the yard light on, we walked them to their car to say goodby.

All talking at once, we'd reached the car when somebody asked, "Where's Grandma?" Then we heard her faint call: "*Help.*" Everybody rushed back and there she was, that little old lady, looking up at us from the bottom of the trench.

"You all right, Ma?" my father asked as he and Uncle Rex pulled her out.

"Oh *my*, yes," she said in her sweet, old-fashioned way. "But I was *ever* so surprised!"

You could have a lot of fun in the hay barn. With the help of a counterweight, you could hang on to the hayfork rope, looped down to the mow where you could grab it, and be lifted, like a trapeze artist, to the peak of the barn.

The counterweight was two or three other kids (my sisters, the Herman kids, visiting cousins, the more the merrier) grasping the rope hanging down from the pulley over the edge of the mow, and jumping off. Down they went, straightening the loop in the rope over the mow, and up you went, some twenty or thirty feet (depending on the amount of hay in the mow), to the top of the barn.

With enough kids you could shoot up almost to the point of bumping your head against the ceiling, and almost lose your grip from the snap of the rope at the top. The girls seemed to enjoy it as much as the boys.

We boys were always upping the ante, though, inventing more daring ways to go up on the rope. One thrill was to sit on the rope, launch from a crosspiece or hay piled against one wall, and swing across to the opposite wall. Once, with a lot of weight launching me, I swung across and up, tipped over backwards, and just managed to hang on as I was lowered to the mow. Another time I hit the opposite wall so hard I rammed my feet through it. That earned me a bawling out from my father as, from the top of our extension ladder leaned against the outside of the barn, he nailed the sprung boards back in place. *Don't do this again!*

And there was the time, trying to match Nicky Herman and another strapping boy, the visiting nephew of our Scandinavian neighbors across the road, I went up hanging by my knees and was snapped off the rope at the top. I was falling, head first, toward the edge of the mow and the wooden floor of the bay, and I suppose might have broken my neck, except that the neighbors' nephew leaned out and caught me. He broke my fall, and we both tumbled to the hard floor without hurting ourselves.

Full into the postwar building boom, my father decided to build a house to sell. If it went all right, he said, we might build another. "We" included me, his twelve-year-old son.

He proposed to build the house on the forty acres below Hamel formerly owned by my grandparents, which we used to rent from them, and which my folks had bought by now. The land fronted on Highway 55 out of Minneapolis, less than twenty miles from the city limits. Hamel would be easy commuting, my father thought, for a guy who worked in town, some young veteran, say, with a G.I. loan and a pregnant wife, both of them looking to live in the country.

Helping my father consisted mostly of standing by as he worked, waiting to fetch this or that tool for him or help hold the board he was cutting or the board he was nailing up. *Stick around,* he'd say, *in case I need you.*

As usual, he was doing things the hard way—the only way he knew how. Every board that went into that house he cut with a handsaw, then nailed into place with a hammer. There must have been power tools by then, but my father never used them. I don't think he trusted them. He trusted his hands, his gnarled, stubby-fingered workman's hands that, forty years later, folded on his chest, would move me to tears as he lay in his casket.

He knew only rough carpentry—picked up, possibly, from Great-Grandpa Williams, who'd been the building contractor in the family. To increase his knowledge—especially about framing—he bought a set of manuals on carpentry (handy, leather-bound little books that had disappeared by the time I might have used them to build our house in British Columbia), and set out, by trial and error and with a little help from me, to build a simple, one-storey bungalow.

It was nearly complete when, on someone's advice, he decided to put a walk-in basement under the house to make it more salable.

That meant raising the house. We used four hydraulic jacks, going from one corner of the house to the other, to lift it inches at a time. Dad did the jacking and I placed blocks of wood under the corners and the house gradually rose, higher and higher, until we could walk under it and hear it groan with every pump of a jack handle and wait for the jacks or the piled blocks to give way and have the whole structure come crashing down on us.

"Careful," Dad said. "The blocks have to be square on top of each other."

Eventually came the labor of digging under the house, with pick and shovel, to level out a floor, and then the business of putting up cement-block walls. Dad did almost all of this work himself (with occasional weekend help from one of my uncles), shoveling dirt into a wheelbarrow to be hauled out from under the house, which stood now as if on stilts, then mixing cement in the same wheelbarrow to bind the blocks he fitted together for the walls and finally to make a floor. About all I did was help carry blocks and then stand around and watch and maybe learn something. *Don't run away, goddammit!*

There were trips to the Hamel lumberyard, for boards or nails or cement; to Fortin's Hardware, for this or that needed tool. Going to either place was like a little vacation from the hard work of building the house. I liked the pungent smell of the stacked lumber in the big shed at the lumberyard. In the hardware store there was the hard smell of metal, and row upon row of interesting tools, household appliances, odd gadgets whose purpose you could only guess at.

When the house didn't sell right away, my mother worried. *I thought we had a lame duck on our hands, but your dad, you know, he was always the optimist.*

Somebody bought it, finally, along with its acre of land between Highway 55 and the Soo Line tracks. The folks made a profit, I think, but not enough to encourage my father to build another house. Instead he sold off the remaining frontage on Highway 55 for others to build houses on, a move he regretted some ten years later when, having quit farming and decided to build a golf course on the Hamel land, he found himself without access to it except for our old entrance across the tracks. That wouldn't do for a public golf course, so he was forced to buy an easement through what had once been his property.

Not long after the house in Hamel had been built, and sold, we got a new hired man. His name was Jack. He was a young war veteran who had worked with my cousin Barbara's husband, Johnny, in the Minneapolis Moline factory (long since closed, its tractors vintage now, collector's items) before being fired for habitual tardiness or failing to show up. As a civilian, Jack was a little lost, Johnny told us. Maybe he'd do better on a farm.

My father, always an easy touch, was willing to give any guy a chance. It was early summer, haying season. We could use the help.

Jack was a small, compact man with a hard, handsome face. He said he was twenty-five, but looked older than that, and had served in the Army in the Pacific. He'd helped liberate the Philippines, he said. That was about all he told us about himself.

"He looks tough," my mother said. "I'll bet he's sinned plenty." She'd invited him to go to church with us that first Sunday after he'd moved into the hired man's room, but he said, "No, thanks. I'm an atheist."

I went upstairs to his room one morning to wake him for chores. I found him lying open-mouthed on his side, drooling into his pillow, his face a cruel mask. He looked positively evil. He'd killed people, I guessed. Known women. Done all those things men did in the service when they didn't believe in God and were away from home.

"Jack," I said. "Time to get up."

"Huhhh?" He jerked awake, staring wildly around as if he didn't know where he was. Then he reached for the pack of cigarettes by his bed and lit one. He was smiling now.

"Whattaya say, Sarge?"

"It's six o'clock," I told him. "Time for chores."

"Be right down."

He sat there smoking, seemingly in no hurry. Outside the door I heard him talking to himself but couldn't make out what he said.

He was a good-enough worker, anyhow, and impressed my father, who started taking a fatherly interest in him. "Hell, we could fix up the old milk house for him," he told my mother. "He could be our permanent hired man!"

The old wooden milk house, now a storage shed, had been replaced by a brick and cement structure attached to the barn. Now, instead of filling the cans inside the barn and letting them stand until after the milking, when they were hauled by stoneboat up to the old milk house for cooling, the cans could be lined up in front of the cooling tank in the new milk house and lowered into the cold water as each was filled. That speeded up the cooling process, reduced the formation of bacteria in the milk, and met the postwar regulations for producing Grade A milk in Minnesota.

"We could plank over the old cooling tank," my father said, "throw a mattress on it for Jack's bed! We could put a stove in there for winter!"

"I don't trust him," my mother said.

"Aw, give the guy a chance. He's been in the war, for chrissake."

Shrugging off my mother's reluctance to spend the money, my father bought Jack a used car, a nifty little Model A roadster with a rumble seat, *just like the one your ma and I had when we was first married*—just the vehicle I wanted for myself someday. Jack was to pay the folks back out of his wages, which my father had upped to one hundred dollars a month, plus room and board. How about twenty-five bucks a month? my father asked. Jack would have the car paid for in no time.

I think my father saw another son in Jack, one better able to help him. I saw him as a mentor, a guy who could teach me things I thought my father couldn't imagine. He picked up on that and played with it. One day he took dice from his duffle bag and showed me how to shoot craps. Another time he told of the fun you could have in places like Manila. Talked of the "gooks" over there and their brown, willing women who would fuck you blind for a couple of dollars. My head filled with gorgeous, hazy visions of exotic female flesh without quite knowing what men and women did with each other. I was a farm boy and knew what animals did. But I was also a Catholic boy who'd been schooled by nuns and uninstructed by his parents, so I had some kind of fuzzy notion about men and women simply lying together and feeling nice and somehow, by a kind of cross-pollination, producing babies.

He took me to play pool with him at the Buckhorn Café in Long Lake a couple of times. Bought me pop and himself a beer. Showed me how he could run the table. We played eight-ball, rotation. I was maybe thirteen by then, just coming into adolescence, and I was as much scared of him as fascinated, thrilled by his tough guy banter and by the fact that he'd been to foreign places and committed mortal sins, apparently without remorse.

One night we were driving home from the Buckhorn into a thunderstorm moving toward us from the west. It was a hot night, but there were cool drafts blwoing through the open windows of Jack's car. Then there were flashes of lightning, claps of thunder. I liked the excitement of an approaching storm, but I saw Jack cringe at every flash, duck at every thunderclap.

Suddenly there was a crackling flash and a simultaneous *ka-boom*, right over our heads. Jack let go of the wheel and ducked under the dashboard with his hands over his head. I just managed to lean over, grab the wheel to keep the car on the road, and switch off the ignition. The car stopped, and we sat there, in the middle of the road with rain pounding on the car's leather top and splattering through the open windows.

"Thanks, buddy," Jack said.

He'd straightened up to sit behind the wheel again. We rolled the windows up and continued to sit there, listening to the rain. Finally Jack said, almost accusingly, "You know what it's *like* under fire?" And then: "Don't tell your pa."

Maybe a week later Jack asked my father whether he'd mind if he ran into Hamel for cigarettes. We were walking up from the barn to the house for supper before the milking. It was a soft summer evening. We were in the middle of haying.

"Sure, go ahead," Dad said.

"Can I go with him?" I asked.

I think my father was about to say yes when Jack said, "No, I'm just going for smokes, Ross. You go on with your pa and eat your supper."

I was standing on the lawn, disappointed because Jack hadn't wanted me along, when he started up his Model A and drove past me out of the yard. I gave him a wan smile and waved. He gave me a funny, distracted look as he turned down the gravel road toward Hamel.

When he didn't come back, my father shrugged. "I suppose he's gone on a binge."

"He's *gone*, period," my mother said. "And not a cent paid back on that car you bought him."

When he didn't return the next day, or the day after, you could see the hurt in my father's eyes.

Then he found his toolbox missing from the shed.

"That no-good son of a bitch!" he said now. "That fucking ungrateful bastard!"

"Oh please, Mervin, don't swear like that." It was my mother's constant plea.

"I'll swear when I goddamn *feel* like it, goddamn it!" And he let go with every swear word he knew, every foul-mouthed exclamation he could think of as he declared, over and over, what he'd do to Jack if he ever came back. "I'll *kill* the cocksucker!"

My mother held her ears. I stood marveling.

"Such language," Ma said, when she could get a word in edgewise.

THE THREE MUSKETEERS AND OTHERS

The 1948 film version of *The Three Musketeers*, starring Gene Kelly as D'Artagnan, swept me up in that romantic story of swordplay and fellowship. I got my closest friends Hal and Dan to join me in forming our own Three Musketeers (I was D'Artagnan, of course; never mind that he was really the fourth musketeer). Our last year at SS. Peter & Paul I had us swaggering through Loretto after school with our buckle overshoes unbuckled and flapping, like the floppy, seventeenth-century boots of the musketeers, and periodically raising our sword arms (*sans* swords) to cross them in the musketeers' pledge: "All for one, and one for all!"

Dan lived in Loretto—we'd been friends since the second grade and my first year in the sister school—and we had a number of adventures together. His father had the Conoco dealership there and delivered fuel to local farmers. His storage tanks stood beside the Soo Line tracks near Dan's house. I climbed with him one day to the top of the gasoline tank, which Dan uncapped and where he introduced me to sniffing gas. It was seductive, if somewhat sickening, and his kid brother, who passed out once and nearly fell off the tank, was especially taken with it, Dan said.

There was the Saturday I rode my bike to Loretto with a sword made from a broom handle and a tin can for a cup guard to engage in a "duel" with my friend.

Dan's sword was simpler but sturdier (as it turned out), just a couple of crossed laths. We saluted each other and began to thrust and parry. *Whack*, and the wooden sword I'd spent all that morning making, broke in two within seconds after a blow from Dan's. We just stared at the stump of it, then broke into crazy laughter.

Up the tracks a ways, past his dad's Conoco tanks, was Loretto's Soo Line water tower. We climbed it that day, or maybe another day. In any case, Dan and I were sitting on top of the thirty-foot wooden tower when I looked down the tracks and saw a man come out of the depot and start walking toward us.

"Oh-oh, here comes the depot agent," said Dan. "He's gonna lick us."

The man didn't seem in any hurry, and I knew why. There was swamp on either side of the tracks here and nowhere to run, it seemed.

"What'll we do?"

"Nothing," Dan said with a goofy smile. "We're trapped."

We climbed down the ladder and stood nervously waiting under the tower for the agent to reach us. Dan seemed resigned to the licking he was going to get. Not me. I plunged down the railroad embankment and into the swamp—into a jungle of cattails and clinging morning glory vines—until I was hidden in the tangle and up to my chest in cold water. Presently I heard the crunch of the man's shoes on the crushed rock between the rails above me and his gruff voice—*C'mere you little shit!*—and then the sound of Dan being spanked and his pitiable cries as if he were being beaten to death. I heard the crunch of the man's shoes again as he walked away. There was a pause. Then Dan called softly, "*Ross! Where are you?*"

I came struggling out of the cattails to Dan's explosion of laughter.

"Didn't he hurt you?"

"Naw. He's caught me a couple of other times. When I yell like that, he thinks he's hurting me and stops."

Another time we climbed Loretto's water tower. The ladder to the top of the hundred-foot metal structure was enclosed by a cage that presumably would give you something to grab onto if you lost your footing; but that didn't prevent my getting dizzy whenever I looked down. Every time I looked down I felt the awful lure of letting myself fall. Dan kept climbing ahead of me, without looking down—he'd done this before—reached the top and disappeared. When I reached the end of the protective ladder and crawled out onto the tank's smooth, slightly rounded top, Dan was sitting on the raised, circular cap in the center of it, grinning at me.

"How *about* this?" he said.

Sitting up there we had a thrilling, bird's-eye view of the whole town and the surrounding countryside. After a while Dan decided to crawl to the edge of the tank to look down at the ball field. It was in a hollow under the hill the tower stood on, so it was a long way down, and when Dan leaned over the slope of the tank I was suddenly sick with fear that he'd slide off. "*Je*sus, Dan!" I said. "Be careful." He laughed.

We were sitting on the cap again when we saw his dad's Conoco truck drive out of town, heading away from us. Then the truck turned around and headed back our way.

"Oh God, he's seen us!" Dan said, and we scrambled down the ladder like a couple of firemen.

But he hadn't seen us, just forgot something and drove back for it, Dan found out later. But *somebody* had seen us because after that the town put a kind of gate at the foot of the ladder and locked it.

Hal, our fellow musketeer, was a farm kid like me, but I'd barely got to know him before he moved with his family to Robbinsdale, a northwest suburb of Minneapolis, where his father worked, I think, in a lumber yard. Then he was back, his folks moved back to their farm and Hal enrolled once more in the Loretto school.

He was different when he came back, much bigger and full of stories about life in the city. He talked—bragged, I thought—of going to Saturday matinees at the Robbinsdale theater to see exciting movie serials that we country hicks had been deprived of: Buck Rogers, Flash Gordon, Tom Mix. I envied his city experience, and scoffed at it.

Maybe that's what started it.

We were walking home from school one day, Hal and the Tyson boys and I. It was late spring, the weather was warm, and it was fun walking home after school now rather than riding in the folks' car with my sisters.

You were kind of bouncing around and irritating me and I started to get on you about various things.

I suppose I was showing some of my envy of Hal, needling him. Then he was needling me. When we reached the crossroad along which he and the Tysons lived, he stayed with me, wanting to fight. I didn't want to fight; he was too big and I was scared of him. He kept after me, though, until we arrived at a place where he could cut across the fields to his place.

At home I whined to my mother about the incident, and at supper that night my father said, "I hear they're picking on you at school. They used to pick on me when I was a kid. Then I learned to box. You wanna lesson?"

We went into the parlor and squared off. He showed me how to make fists with my thumbs curled over my knuckles so they wouldn't break when you hit somebody. He showed me how to present myself, hunched forward, head down, forearms up, my left slightly ahead of my right. You jabbed with your left, kept your right up to guard your face. Jab, jab. Then a right cross. "Not a round-house right, you'll only hurt your arm and leave yourself open. Keep your arms in tight and your head down. Get in close." He showed me some footwork. "Keep moving! Keep jabbing! Stay on the balls of your feet!"

He danced around to show me. "Okay," he said. "Now try and hit me."

I took a poke at him.

"Naw, naw. Try and *hit* me. Go ahead."

I swung at him, hard, and he blocked it, then tapped me on the chin.

"See? I coulda floored you then. You were wide open."

He showed me how to bob and weave, how to feint, how to set a guy up for a right cross, for the old one-two. He taught me how to keep balanced, how to get inside your opponent's reach, how to body punch, how to clinch, how to keep your face and eyes impassive so as not to telegraph your punches.

That's when I heard he'd learned to box in a Minneapolis gym and had fought a couple of amateur bouts as a featherweight. He'd quit after seeing too many punch-drunk, washed-up fighters working out in the gym.

At school the next day, at the start of noon recess, I swallowed my fear and stopped Hal as he was leaving the classroom. I said I was ready to fight him now.

"Well . . . okay," he said, a little leery. "But remember—you started it."

We squared off in front of the empty desks. He swung at me and I ducked, got in close, and caught the point of Hal's nose with a blind uppercut. He spun around and bumped his head against a radiator. When he turned back to me, he was holding his nose and blood was gushing through his fingers.

"That's enough," he said.

We were friends after that.

The Tysons were a family of four boys and a girl. I took to visiting them on Saturdays during the school year, walking a couple of miles across the fields to their place, and pretty soon it felt like a second home. I was closest to Cal, who was a year younger than me. The girl, Elizabeth, was a year older than me, but she was pretty and of course I developed a crush on her.

They were great drinkers of coffee. In the midst of their workday they took an afternoon break—something my folks never did, unless it was to take a nap—during which they gobbled sweets and drank cups of the bitter coffee the missus kept simmering in a gallon pot on the kitchen range.

Their house was a nineteenth-century clapboard like ours, heated not by a furnace in the basement, as ours was, but by the kitchen range and stoves in each of the other rooms downstairs. It must have been a job just tending all those stoves.

They showed me their collection of Indian artifacts, a cigar box full of arrowheads, spear points, stone scrapers, found in their fields over the years.

"We think there was an Indian camp around here in the olden days," said Cliff, the oldest of the Tyson boys. He was already in high school, and hardly associated with the rest of us.

    I envied their collection, and I envied the fact that all the boys had guns. Little Herbie had a Red Ryder BB gun. Cal had his own .22 rifle. Willie had a .22 and a sixteen-gauge shotgun, and one day he asked if I wanted to shoot it.

    "Yeah!"

    The Tyson boys stood back grinning as I raised the long gun, barely able to lift it, and aimed at a tin can Willie had put on a fencepost. *Bang* and the gun's kick jarred the snot out of my nose and practically knocked me off my feet. Everybody laughed, and I knew I'd passed some kind of initiation.

    I'd set out to visit the Tysons after morning chores on a Saturday, then head back to reach home in time for evening chores. I grew to like those lonely treks to and from their place, especially in winter and especially homeward bound in the late afternoons, when I trudged through the snow across the darkening fields and into stands of bare, silent trees. Often I carried an exciting bundle of borrowed magazines, *True, Argosy, Outdoor Life,* all of which the Tysons subscribed to. The snow was soft in the woods but was swept hard out in the open by the winter wind, so you could walk on it. In the dusk I might see a snowshoe hare in its white winter coat, like a ghost against the ghostly landscape.

    There was an adventurous shortcut to the Tysons'. It was through the woods at the far end of my aunt Betty and uncle Gerry's farm that descended into a tamarack swamp and ended below the Tyson farm where a creek came out of the swamp and there was a rotting wooden bridge across it. That woods and swamp was a jungle you could get lost in, and you could only cross it in late summer or fall, or in the winter, because earlier in the year it was mostly underwater. When it was dry or frozen, though, you could head straight through to the Tyson place.

Crossing it one warm winter day, wading through the soft snow under the trees, I came upon a hump of high ground, a hidden island in the swamp. I saw movement, and a wild-looking man appeared. He tramped down off the island in snowshoes and told me gruffly to stay away from his traps. The Tysons, when I mentioned seeing him, said he was an old hermit, possibly crazy, who trapped mink and fox and other fur-bearing animals for a living.

The Tyson farm was nestled under the brow of a hill by their own little pothole lake. The boys took me fishing in the lake, one of them rowing the boat as I trolled for northern pike. Caught one! For fun, we took their dog, a hound of some kind, out to the middle of the lake and threw him overboard, then watched him swim resolutely to shore.

The pasture along their lake rose steeply into a hill that the Tysons slid down in the winter on "tin toboggans," sheets of corrugated roofing that, with a bent-up end, worked wonderfully, scooting you down the almost perpendicular slope to whack through the cattails along the shore of the lake and skim out over the ice. You didn't kneel as on a regular toboggan but sat and leaned back on your hands, steering with them when necessary, sometimes spinning out of control and flying off the sheet of tin.

One day when we were out sliding there was another visitor at the Tysons', a kid from Loretto named Rudy, who was one of those unfortunates among school children like the gamma member of a pack of wolves. He was the butt of jokes, and the Tysons especially teased him unmercifully. Having been teased and the butt of jokes myself, I felt sorry for Rudy. But then he seemed to ask for it by his craving for acceptance, his tail-between-his-legs hanging around alpha kids like the Tysons. It didn't help matters that, besides wearing glasses, he had a whiney voice and was a bit of a sissy.

Before the Tyson hill dropped abruptly to the lake, there was a long, more gradual slope you could use to build up your speed before sailing over what amounted to a drop-off. Taking turns on the tin toboggans, we started competing, going farther and farther up the hill to start our individual runs.

Finally I decided to start from the very top of the hill. Rudy, no doubt wanting to prove something too, asked to ride with me. Together we climbed to the top, then turned around to see the Tyson boys kicking at the snow along the edge of the drop-off.

"Get out of the way!" I called stupidly.

They jumped out of the way as Rudy and I started down. Approaching the drop-off, at increasing speed, we saw what the Tysons had done—made a ramp, those rascals, that caused Rudy and me to sail out into midair. We must have been some forty feet above the ground for a second or two, and drifted down like ski jumpers riding a single ski to land with a jolting *whump* at almost the bottom of the slope and go skittering across the snow. Rudy lost his glasses. I'd turned around somehow during the flight and landed facing backwards; the jar, just as shooting Willie's shotgun had done, knocked the snot out of my nose.

The Tysons' gleeful laughter pelted down on us like snowballs.

The Tysons had another dog, an old part-collie they said could beat any dog around. I said I had a dog too that was a pretty good fighter.

"Bring him over," Willie said.

This was maybe a year or two after my dog Shep had chased our cattle through barbed wire, ripping their udders, and we'd both been spared my father's wrath. I figured Shep could fight because I'd seen him kill one of our cats. It had been a needy, purring young thing that stayed around the house, instead of the barn, and she and Shep seemed nicely tolerant of each other. One day, though, for some reason it hissed and arched its back at Shep when he came up on the porch with me. On impulse I said, "Sick'm."

Shep didn't do anything at first, just looked intently at the cat, which stood arched and frightened in a corner. Then, in a single, quick movement, he snapped the cat up in his jaws and shook the life out of it. I was flooded with regret, then filled with fear, as I carried the body behind the house and buried it before anyone saw what Shep and I had done.

The next Saturday I walked with Shep to the Tyson place. The Tyson boys and I gathered around as the two dogs circled each other, bristling. We sicked them on each other.

They went at it, snarling, biting, tearing at each other's snouts and necks, spinning and rearing, trying to pull each other down. It was terribly exciting and faintly unnerving to watch. The air filled with their frantic growls, their high-pitched whines. You could hear their jaws snapping.

"Look at'm go!" Willie Tyson cried. "Get'm, Rex! Kill'm!"

It looked like an even match. Finally the dogs fought to a standstill, stood on their hind legs and just leaned on each other, their jaws dripping saliva and blood, too tired to continue.

"Aw, they're done," said Willie. "I'd call it a draw."

The two dogs had dropped to the ground and moved away from each other. They were done, all right, but something made me say "Sick'm" again to Shep and he went at the Tysons' dog, got him down, and started chewing at his neck as he lay still, taking it. Willie said, "Call him off! He's gonna kill our Rex!"

We were silent then. Finally Willie said, "Your dog's a lot younger. But old Rex almost beat him, didn't he?"

Rex lay panting on the ground. Shep collapsed, almost beside him.

"Let's not fight them anymore," one of the Tyson boys said.

3

The Woods and the Farm

BOYHOOD EDEN

One near-freezing night in the spring of 1947, when I was twelve years old and feeling the first faint twinges of adolescence, I sat bundled up on a lawn chair outside our farmhouse listening, overcome by an almost painful rapture, to the shrieking chorus of spring peepers.

Thousands, millions of the tiny frogs you almost never saw, only heard their shrill singing, were calling from the big swamp beyond our neighbor's farm across the road and from our swamp pasture—calling, it seemed, to me alone. I was in love, you see, in love with nature as if nature were a girl.

Of course nature on a farm is mostly domesticated. Farming involves the keeping of animals and the growing of crops. Simply by living on a farm you witness birth and death, see procreation, experience the cycle of the seasons. You absorb these things as part of your world, during the course of your work or play. At the same time you're surrounded by the part of nature that isn't domesticated, that's wild. It's another world altogether, related to yet detached from the domesticated world of the farm. Two worlds then: "the woods" and "the farm."

The woods for me included Elder's Forty with its wooded pasture and other upland and lowland remnants of the native forest around the farm, tree-covered ridges and wooded ravines, wild stretches of marsh and lakeshore—"wasteland," in short, to a farmer, land too steep or broken or too wet for cultivation.

It was in these little patches of wilderness, in solitary explorations that began with timid, close-to-home forays and widened as I grew older and ventured farther, where I lived out my boyish fantasies, pretended to be an Indian or a wilderness scout or a naked wild boy.

I was a boy naturalist. Observing, reading everything I could find about birds and animals (issues of *National Geographic* out of the SS. Peter & Paul library were a handy source of information—despite what my friends and maybe the nuns suspected, I wasn't only interested in the magazine's pictures of bare-breasted native women), I got so I could identify most of the wildlife around me. Birds especially, before pesticides and wholesale destruction of their habitat depleted their numbers, seemed to fill the countryside.

Going to get the cows, in the freshness of a summer morning, I'd hear the doleful, somehow peaceful cooing of mourning doves. Grackles, both the bronzed and purple varieties, darkly glistening, stalked the grass beside the cowpath like miniature crows, their bright yellow eyes regarding me as I passed. Among the cattle I'd find brown-headed cowbirds, walking behind the grazing animals, feeding on the bugs stirred up by their hooves. Cowbirds laid their eggs in the nests of other birds, I'd read, like cuckoos.

From the fields I'd hear the call of the killdeer, *kill-dee, kill-dee,* as it's transliterated in the bird books. They nested on the ground in the fields, and when you got too close the mother bird went into her broken-wing act, showing the bright orange of her rump as she led you away from her nest. You could never catch her.

From sloughs came the pleasant, distinctive call of red-winged blackbirds: a *check,* then the tripping, three-note *O-ka-re* (another transliteration), the last note sounding like a rusty spring uncoiling. Occasionally, among the redwings, I'd see the larger yellow-headed blackbird, which was more common farther west, while from fenceposts came the tinkling song of meadowlarks. Meadowlarks flew like miniature pheasants, alternately flapping and gliding. Goldfinches had a looping flight and chattered excitedly in the air. Doves flew rapidly, on whistling wings. And in the early summer there were bobolinks, flying up to sing their bubbling song, then falling back into the grass. By July they'd be South again, where they were called rice birds and swarmed in huge flocks and were shot as pests, I'd read. In the West Indies, where they wintered, they were called butter birds and killed for food.

Above the fields red-tailed hawks, and other buteos, soared and tilted in the thermals. I'd see sparrow hawks, more properly called kestrels, the smallest falcon, perched on fenceposts and telephone or electric wires, or hovering over a hayfield, on the lookout for mice or grasshoppers. Sometimes I'd see a marsh hawk, or harrier, coursing low over the fields like a hunting dog with wings. What I never saw out in the open and only seldom in the woods was an accipiter, either the long-tailed, short-winged cooper's hawk or the smaller sharp-shinned hawk, both quick, artful dodgers among the trees after songbirds.

In the wet grass were hopping frogs and the slithering garter snakes that preyed on them. You might come upon a frog mesmerized by a snake, about to be grasped, or a snake in the act of ingesting a frog, unhinged jaws working, the frog disappearing headfirst into the snake's bulging neck. And in the sloughs and marshes were all the waterfowl: ducks and coots and snipe and bitterns and great blue herons. The wild creatures went instinctively, absorbedly about their business, it struck me one day, feeding themselves, mating, protecting their territory, absolutely *of* the natural world, *their* world. It was a world I wanted to know, sink into, become one with—I wanted to become a fellow creature of the wild.

I didn't study the vegetation but knew the crops we planted, some of the native grasses, the common trees, trees I walked under in the woods. The deep woods, I learned, were remnants of what the first white settlers in Minnesota knew as the Big Woods, and held stands of oak and maple, basswood and ironwood, and along the water courses elms, cottonwoods and willows. I knew the box elders in our yard and knew from their winged seeds that they were a species of maple. I wondered at the big willow that shaded the outhouse, displaced and yet so healthy on that high ground, until a hired man suggested it drew moisture and nutrients from the privy stew.

Like that outhouse willow, throughout my childhood I absorbed what I now realize were the riches of my environment; gloried in the hot, lush summers, endured the cold, bleak winters, fixed in my mind the gently rolling, pastoral landscape of south-central Minnesota, the look of its mixed prairie and hardwood forest after a hundred years of white settlement—embraced its fields and woods, its bogs and marshes and glacial pothole lakes, so that even now, after more than thirty years in the mountains of southeast British Columbia, it remains my home country. It's the country of my heart, as D. H. Lawrence said of his rural England; it's where, in the woods near our old farm, I'd like my ashes to be scattered someday.

I longed to have some bird or animal from the wild for a pet—wanted to capture and tame it so as to achieve its acceptance, understand some of its mystery. I saw *The Yearling,* the 1946 film based on Marjorie Kinnan Rawlings's wonderful 1938 novel (I read the book after seeing the movie), and passionately identified with the boy Jody, his closeness to nature and his love for the deer he raised from a fawn. I too wanted a fawn to raise, without having to shoot it when it grew up, as Jody had to, after it kept ruining his mother's garden.

But apparently there were no deer anymore in our region—not then—so finding a fawn in the woods, as Jody did, wasn't likely. Later, when I was in my teens, deer began to reappear, after having been driven north by pioneer settlement and uncontrolled hunting, so that by the fall of 1950 there was actually a short season for whitetail deer in our area—shotgun slugs only, no thirty-caliber rifles that could send a bullet through a cow, or a farmer, a mile away. My father and I, along with most of our neighbors, spent a day or two hunting the woods. I didn't see anything to shoot at (I had my uncle Isadore's long tom shotgun; it stood as tall as I was), but John Herman got a deer, and my father, the afternoon *before* the season started, shot a nice buck on Elder's Forty, only to find it gone when we went in the dark to fetch it home after evening chores. Whether a bow hunter found it (bow hunting had started the day before), or it managed to get up and walk away, we never knew. All we found was blood on the ground where it had fallen.

One spring day I caught a young crow in the woods. I heard its hoarse cry and found it on a low branch, not yet able to fly after apparently falling out of its nest. I chased after it, caught it, threw my jacket over it and carried it home.

Crows were smart and elusive, bold raiders of the fields (you could hardly draw a bead on one with a gun; they knew a gun from a hoe, it was said), and I was tickled to have one. But it died after a few days. I might have fed it raw hamburger, somebody told me later, or ground corn and oats softened with milk; instead

I poked kernels of raw corn down its gullet, which it threw up, and so starved it to death.

I did better with a pigeon. There were pigeons in the grandfolks' barn in Hamel, and early the next spring when the barn was empty except for a dusty layer of last year's hay in the mow and stuffy with the acrid smell of pigeon droppings, I climbed to a nest under an eave and found eggs in it. I kept checking until the eggs had hatched and the nest was crowded with naked and ugly squabs, like little dinosaurs. I waited until the largest had the fuzzy start of feathers, then took it home and put it in the lamp-heated brooder house with our baby chicks.

I named him Peter: Peter Pigeon. Every day, before and after school, I fed him by hand with morsels of bread soaked in milk, then with grains of wheat. Pretty soon he thought I was his mother and would come running toward me when I entered the brooder house, peeping excitedly, his sprouting wings outstretched. Soon he was eating out of my hand and eventually learned to eat out of the chicks' feeder. He grew up thinking he was a chicken, and followed our free-ranging birds around the farm, pecking with them at the ground. I wanted him to fly. I'd pick him up and throw him into the air, but he'd only settle on the ground again.

He grew scared of me finally and did fly, short escape flights when I tried to catch him. Then one day something remarkable happened. A flock of pigeons winged by overhead, and he flew up after them. He had to struggle to reach them, his rapid, laborious wingbeats no match for his wild brothers' clipped, easy flight. But he managed to climb to their altitude and catch up with them. Then he was gone.

Among the books I was reading then (*consuming* was more like it), apart from *Classics Illustrated* comics, field guides, "biographies" of birds and animals, westerns, historical fiction, anything to do with wilderness, Indians, woodsmen and mountain men, was a series of novels for juveniles, written at the turn of the last century and set on the frontier around the time of the American Revolution. The series was called *The Young Trailers*, and its hero, named Henry Ware, was patently modelled on Daniel Boone, I realized; that he also represented the author's and his period's racist assumptions, I wouldn't know until years later. Henry Ware is a white man who beats the red man at his own game. He's a Nietzschean superman of the type Jack London was creating about the same time. I completely identified with him. I wanted to be him.

After discovering the series in the library at SS. Peter & Paul, I read all eight books, year after year, until I left the school. My friend Hal read the series just as avidly as I did, both of us taken by Henry Ware but also delighting in his faithful companions, Paul Cotter, "Shif'less" Sol Hyde, "Silent" Tom Ross, and "Long" Jim Hart. Together they formed a company of scouts, "woods runners," that ranged the continent's forested wilderness east of the Mississippi in the latter years of the eighteenth century, spying on the British and their Indian allies and thus aiding American frontier settlers. That they were only helping to destroy an Eden in which they were as much at home as its native inhabitants didn't dawn on me then—nor, it seemed, did it occur to Henry Ware and his friends or to their creator. Reading *The Young Trailers* made me long to have lived in their time, to have known, as they did, the vast, unspoiled tracts of North America in those lost, exciting days before I was born.

Eventually I discovered Mark Twain, whose books were noticeably absent from the school library. But I found *The Adventures of Tom Sawyer*, as earlier I'd found *The Golden Book* anthologies, in one of Aunt Lelah's old steamer trunks in our upstairs "junk" room. I read and reread *Tom Sawyer,* fixing on chapters thirteen through sixteen in which Tom and Huck Finn and Joe Harper run away to camp on Jackson's Island in the Mississippi—read those chapters over and over until they fell out of the book, then carried the detached pages around in a pocket of my jeans or bib overalls to read in spare moments as you might some inspirational tract.

Tom and Huck gave me a style. Inspired by the illustrations in the book, I began to go around barefoot, in rolled-up jeans, which angered my father (*How the hell can you work in bare feet?*) and so toughened my soles that pretty soon I could walk barefoot across a stubble field without cringing.

Tom Sawyer soon led to *Adventures of Huckleberry Finn*, which I found in the little Hamel branch of the Hennepin Country Library. I realized it was a much better book than *Tom Sawyer* because of its "voice" and its lovely descriptions of Huck floating down the Mississippi on a raft with his friend Jim. That Jim is black and a runaway slave, and that the book's "message" has to do with the evils of slavery and the brotherhood of man, hardly registered on me as a northern white boy with no knowledge of slavery and who knew blacks only from the sight of them out the window of our car as we drove through north Minneapolis. Like Tom and Huck, like my folks and our neighbors, I spoke then of "niggers" without shame or embarrassment.

I was full into my barefoot Tom and Huck days when the Nordstroms came to our farm. They were a young couple my folks found in the "Gateway" on Washington Avenue in Minneapolis, where farm workers could be hired. Wes, who'd been 4-F during the war and so didn't serve because of some physical or even mental disability, was a blue-eyed and sandy-haired young fellow with a funny habit of clearing his throat before he said anything; a Norwegian from Little Falls, Minnesota.

"Ahem . . . that's Charles Lindbergh's hometown," he'd tell you, as if some of that famous aviator's greatness had rubbed off on him.

Rose, his "wife" (their partnership was common law, or what my mother called "living in sin"), was a short, dark, bosomy Italian with a foul mouth and a raunchy sense of humor. "C'mere, you little shit," she'd say to her two-year old daughter, who toddled about the house stinking, carrying a load in her diapers. Then Rose would laugh at her own crudity. "Shit, shit, shit," she'd say. "Whattaya think of that?" Her coarseness only increased her attractiveness somehow; apparently she'd never lost that childish delight in talking dirty. Even my mother had to laugh.

Wes, too, was "immature," as my mother said, and seemed more comfortable with me than with my father, though he was a good-enough worker, Dad thought. He spent a lot of his spare time with me, treating me like his kid brother. He was in fact the buddy my hardworking father could never be for me.

Wes took me fishing on summer evenings after chores. He'd look at me and wink. "Think they'll be bitin' tonight, Ross?" Then he'd turn to my father. "Okay if I take him fishing with me, Merv?"

"Yah, go ahead," my father would say.

"What about you?"

"Naw. You go ahead." My father was a hunter not a fisherman. His one indulgence was to take time off in the fall to go pheasant hunting in the fields and duck hunting in the marshes.

Wes would send me to the straw pile to dig for worms and grubs while he and my father let the cows out to pasture and washed up the milking utensils. Then, his jalopy loaded with our cane poles and tackle boxes, we drove to one of the lakes. We tried Spurzem's and Independence, finally settled on Sarah as the best lake for sunnies and crappies. It was idyllic, sitting in a rowboat on the still lake on those long summer evenings, watching your bobber as the sun went down and the air turned chilly. Toward dark, we'd row ashore with our string of a dozen or so fish (we hardly ever got skunked) and drive home to show off our catch. My mother started prompting us to go fishing. She could eat a whole platter of sunfish, fried with cornmeal, herself.

One night on Lake Sarah, Wes pulled a pack of cigarettes from his shirt pocket, shook out a couple, and offered one to me.

"Go ahead. I won't tell your folks. You must be smoking by now. *I* was at your age."

"Ah . . . thanks," I admitted.

Sure, I was smoking already, following Tom and Huck's example, sneaking it at home or with other boys at school during recess behind the line of evergreens along one end of the schoolyard, or in the boys' can, or, most adventurously, up in the church belfry, squatting in pigeon shit. Real tobacco, though, was hard to come by for a twelve- or thirteen-year-old, but I learned to watch for butts along the sidewalks in Hamel or Loretto. I'd tear off the germ-laden part where the guy's mouth had been, and smoke the rest of the butt; or, after I'd made a corncob pipe (with a weed stem, just like Huck Finn's!), I'd tear off the paper and save the tobacco in a Bull Durham cloth pouch I'd found and smoke it in my pipe later. Then there were the tobacco substitutes: coffee grounds, crushed tea leaves, dried corn silk, and a weed found in the pastures called Indian tobacco.

There were also lily pad stems, which dried into porous, crooked sticks that could be smoked like cigars. That was an awful, tongue-stinging experience. Smoking generally was an unpleasant experience, and inhaling nauseated me. But I kept gamely at it for a year or so.

The next time we were out fishing, Wes offered me one of his Luckies. I was ready for him.

"No, thanks," I sniffed. "I've got my own."

And I took out a pack of Chesterfields, which some older kid had bought for me, and pulled one out. Wes laughed, then gave me a light. Then we puffed away in comfortable silence together, like two old fishing cronies—Huck and Jim on their raft.

Roaming the wild places with my Stevens bolt-action .22 rifle in the crook of my arm, pretending it was a Kentucky long rifle (actually it was my uncle Billy's gun, left on our farm while he was off serving in the Navy), I pretended to be Daniel Boone or Henry Ware.

"Going out in the woods!" I'd call to my mother. "Be home for chores!"

Most often, because it was closest, I went to Elder's Forty. I stalked through the trees above the pasture's creek bottom or walked the cowpath along the creek to where it climbed over a knoll and came out into the big clearing that provided most of the pasture on the property. Across the clearing was a pond, green with duckweed and algae, where a pair of wood ducks might rise at my approach. Past the pond was the barbed wire of our line fence and the wooded back end of our neighbors the Hermans' original claim. In the Hermans' woods, where they started to descend into a swamp, you could climb a tree and see across the brush and willow-tangled swamp to the open cattail marsh that stretched the three miles or so to Hamel. From the tree's top branches you could see the town's church steeple, sticking up above the trees on the hill where the town stood.

Spurzem's Lake, though, was where I lived out most successfully, most happily, my boyhood fantasies. It was my Jackson's Island, my boyhood Eden.

I'd pack a cotton flour sack with a hunk of bacon, a potato, salt and pepper, a carrot maybe, or an apple or orange, tea or instant cocoa, matches. Then, with the sack hanging from the handlebars of my bike, my rod and reel strapped to my back, my army-surplus cooking kit under one shoulder and my hunting knife and army-surplus canteen at my belt, I'd pedal over the rolling countryside, for a mile or so, to Spurzem's Lake.

I'd hide my bike in the slough grass beside the gravel road and walk in along the marsh that bordered most of the lake to where the shore was firm. Here was pasture, part of the old Spurzem farm (deserted in this period), and a gully emptying into the lake where you had to wade far out on the silty bottom, an almost quicksand mixture washed down from the fields, to reach water deep enough for swimming. Afterwards you pulled slug-like leeches off your legs—bloodsuckers—that I romanticized into tropical creatures of the Amazon or the Congo somehow migrated to this Minnesota lake.

Past this place, across a fence into pasture owned by our neighbors the Jensens, you could walk between a ridge of high ground and the marshy lakeshore to a rotting plank bridge over the stream leading out of the lake. Cattle once had crossed here to reach what constituted a small island, a wooded and grassy knoll surrounded by lake and marsh.

The cove below the knoll had a hard sandy bottom. The water, though, was hemmed in by lakeweed and too shallow for swimming. But there were turtle eggs in the sand—the oval ones were mud turtles', the round ones snappers'—and I cooked and ate some once, just like the boys on Jackson's Island.

The best fishing was off the rotting bridge. Once, sitting there, I watched a mink swim toward me—until I moved or it caught my scent and flipped under the creek's surface. Blackbirds sang among the cattails. The sun was warm on my back, and, looking down, I saw fish in the shadowed water under the bridge.

Often I'd simply set a line off the bridge and go exploring. When I came back, there would be a scaleless yellow-bellied bullhead or a white-bellied catfish on my line, which invariably had swallowed the hook so I had to pull it out with a pliers. Then I skinned the fish with a pliers and gutted it with my knife, being careful of its spiny whiskers, its "stingers." I'd make a small "Indian" fire of sticks and fry the bacon. Then I'd cut up the potato and fry it in the grease. Finally, I'd fry the fish and have a feast.

At Spurzem's Lake I could walk around naked all day, sometimes playing with myself, often sustaining, all innocent of masturbation, a pleasantly aching, bone-hard erection.

Pretending to be Kianga.

Kianga was my definitive alter ego by then, a composite of Mowgli from Kipling's *Jungle Book*, Edgar Rice Burroughs's Tarzan, and naked Amazon Indians and Australian aborigines out of the *National Geographic*. He was my wild boy self, naked and free, at home in the natural world.

I didn't know then that the Jensens, on whose land I was trespassing, were observing me. *We used to see you, running around naked in our pasture by the lake. You were a strange little boy.*

One evening I was walking up from the barn after doing my chores. It was in the late fall, and fully dark. As I neared the house, I saw a figure in the unshaded parlor window.

It was Rose, bent over naked in our metal washtub, breasts swaying, scrubbing at her knees with a soapy cloth.

I drew as close as I dared, just out of the light from the window, my heart pounding, and simply stared—stared at Rose's adult, unclothed female body, at the lushness and fullness of her, the rounded, thrilling contours of her womanliness that held, I must have known even then, much of the beauty and power, the mystery, of nature. Feeling guilty and extremely lucky, I stood there outside her window, shivering with the excitement and the cold, until, too soon, Rose put on her bathrobe and left the room.

I waited before going into the house.

Later, in the privacy of the pantry, I was taking a bath in the same washtub she had used when Rose barged in to get something off the shelves.

"Hey!" I said.

"Haw!" she countered. "You ain't got nothin' *I* ain't seen before."

"Yeah? Oh yeah?" I couldn't stop myself from calling after her. "Well, *you* ain't got nothin' *I* haven't seen before either!"

Her head popped back in the doorway, a smirk on her face.

"You *seen* me tonight, didn't you."

I smirked back at her.

"Humph! That'll teach me to pull the shades when I take a bath."

It was a family joke for a while. "The Peeping Tom," Wes called me, and everybody laughed. But that teasing humor, that easy release of uneasiness, didn't erase the thrilling, disturbing image I now had of Rose—Rose in all her naked glory.

Changes

February 1949, and my mother's sixth child is born. She is named Nancy, in memory of the first Nancy, born in 1937 when we were still living in Minneapolis and who died at six weeks of whooping cough. My sister Joyce and I had whooping cough at the same time, and I might have died of it too, my mother's said, the day she was downtown with me and I began coughing and choking, turned blue, but then was revived by a traffic cop. Little Nancy, though, was too young to survive. The name itself, a favorite of my mother's, later would seem to her to have a curse on it.

That spring my father decided to quit dairy farming. The price of milk had dropped after the war, as predicted, and dairy barns were standing empty as farmers looked for easier money, according to the *Farm Journal,* to which my father subscribed. The cost of living was rising, while a farmer's share of the food dollar was fifty cents less than during the war.

My father talked of raising beef cattle—plowing up pasture, growing more hay, maybe buying a baler. Building a feed lot, buying some steers.

My mother had cousins, stuck, as it seemed to her (and to me, when we visited there), out on the absolutely flat, featureless, though rich-soiled prairie outside Grand Forks, North Dakota. They raised beef and grew wheat, and seemed to work less hard and make a lot more money than we did. *They had their winters off!* Oh, they had to put up hay for the cattle and get their wheat in, and there was the worry that went with one-crop farming; but once the grain was harvested, they were mostly done for the year. *No cows to milk!* Just the steers to feed in the lot, and one man could do that. The rest could take off for a while. Cousin Johnny from Grand Forks came to stay with us for a week one time around Christmas. Often his folks showed up and made the rounds of the relatives. After a good year, they might even go to Florida or California. Imagine that! Instead of grain, Dad said, we could grow more alfalfa to feed our stock and have bales of it left over to sell to ranchers out west.

So, despite my mother's cautious reluctance, the folks sold our cows—not at auction, animal by animal, but the whole milking herd to a single buyer, some big farmer to the south or west of us.

Two big semis pulled into our yard one gray morning when the snow was gone and the first buds were on the trees. I was just getting up for school. Dad had gotten up earlier than usual to be sure to have the cows milked before the trucks arrived. I'd been excused that morning from cleaning the barn, which had been my job before and after school for the last year, since we didn't have a hired man anymore. I was it, I was told. Dad must have cleaned the barn after the cows were gone.

Through my opened upstairs window I heard the cattle being prodded up the chute into the slatted trailers. There were close to thirty of them, not counting youngstock, cows come fresh and milking, heifers with their first calves in them, all more or less purebred but unregistered Holsteins built up over the last nine years. I knew each obstinate, bovine one of them, their individual, black-and-white markings, their individual characters, since I'd weaned most of them as suckling calves, introduced them to the pail and to solid feed, and now they were going.

We were keeping the young stock, the calves and unbred heifers, as insurance until the folks decided definitely what to do. The young stock could be sold later, or shipped to South St. Paul for beef. *No more cows to milk.*

It seemed the end of one kind of life and the start of another. Sure, it was just another kind of farming we were going into, and I knew even then that I would never make a farmer, though my father wanted, *expected* me to take over someday. Still, it was as if we were going out of farming altogether. I was a little sick with the excitement.

I was fourteen that spring. In June I would graduate from SS. Peter & Paul in Loretto, and in the fall go to the public high school in Mound. Mound was a den of iniquity, according to the nuns and Father Michael, but I was thinking about the girls there and what I'd heard they did—that they smoked and even necked with their boyfriends in the school parking lot during the noon hour. That was something else to be excited about. High school itself would be another life for me.

At the supper table that night, my father said, "God, the farm feels empty all of a sudden."

"Just think!" my mother said. "No milking tonight!"

"Makes me nervous," Dad said. "Whattaya think, Ruthie? We do the right thing?"

"I don't know," Ma said. "Personally, I don't miss the cows."

I didn't miss them either, but I kept quiet. Joyce said, "It's almost like we aren't farmers anymore."

"Oh, we're still farmers," Dad told her. "Don't think we ain't."

Right after school let out, in the week or so before first-crop haying started, the folks, baby Nancy, and my sisters and I crammed into our 1940 Dodge with its chrome hood ornament in the form of a bighorn sheep and headed for Lima, Ohio. The twins, I think, stayed with the grandfolks in Hamel.

We were going to visit Dad's relatives in Lima, members of his father's family, whom he'd met as a child in Canada but hardly remembered, and whom none of the rest of us had ever seen. For my father, it must have been something like a search for *his* father. He would see where Will Klatte grew up, where he'd set out from for Canada to become a CPR engineer, only to die ten years later in a train wreck.

By now the folks were undecided about going into beef. They would hold off and think about it. Anyway, my father had bought a baler, and to help pay for it he and I would do "custom" work when we got back, baling for neighbors tired of putting up loose hay and ready to have it chopped, compressed, and tied into blocks of fodder you could pile into your barn like bricks. Meanwhile, we were off on our first vacation.

The folks sat up front in the car, with Dad driving and little Nancy cuddled in my mother's arms. That left my two sisters and me to squabble in the back seat.

"Settle down back there," my father growled.

"You kids'll have to get along," my mother added. "Look out the window. Isn't there some game you can play?"

We watched for Burma Shave signs, shouting out the phrases on the separate little signs that formed slogans as we passed. We watched for cars with license plates from other states. And we watched the flat or rolling countryside of the Midwest slide by, the wooded bluffs and pastured dells of Wisconsin, the level corn fields of Illinois, then the steel and concrete towers, thronged sidewalks, congested, hazy streets of Chicago as we drove through. The big city! I was to see it again on Navy boot liberty six years later, and six years after that come to live and work here as a journalist.

Dad made the two-day drive almost without stopping. There were brief stops, of course, for gas, for calls of nature (or to change Nancy's diapers), for the folks to buy food in country grocery stores along the way: cheese and crackers, salami, bread and peanut butter, fruit, pop, quarts of milk. We ate in the car. We peed in the car too, my sisters and me, using a coffee can and looking away from the person doing it. Afterwards it was my job to empty the can into the flow of air past the car window without spraying myself. Once we heard honking, and looked back to see the car following us with its wipers working and grinning faces behind the windshield. That was the big joke of the trip.

We stayed in a motel, I think, the first night out, somewhere in Wisconsin near the Illinois border, then drove through Chicago and across Indiana the next day. Late that afternoon we crossed into northeastern Ohio and found it rolling and wooded, like back home, and in the evening pulled into Lima and found the house of one of Dad's long-lost relatives. We all must have stayed there that night, the folks and baby Nancy in a guest room, my sisters and I on couches or the living room floor. The next day there was a gathering of our Lima relatives, after which the folks stayed put, I think, and we kids were "farmed out." I stayed with the Wilsons.

Herb Wilson had married the daughter of one of Will Klatte's sisters. He was a railroad engineer, as Will Klatte had been, and was on a run while I was there. I never saw him.

That first morning in the Wilson house I woke to find everybody else still sleeping. I'd slept in—it was after eight by the clock downstairs—and still nobody else was up. I nosed about the sunlit house, gazed out the front window at a tree-lined street. I was hungry but too shy to help myself in the kitchen. I went out, walked around the block. Returned to the house and sat looking through magazines in the living room. I was a country kid, bored and alienated, surrounded by streets and sidewalks, unnatural rows of planted trees, all those houses, my early childhood years in Minneapolis forgotten.

About ten, I heard movements upstairs. Mrs. Wilson came down. "Oh, Ross! My gosh, how long have you been up?"

"A little while."

"Farm boy, huh? You have chores in the morning?"

"Yuh."

"Well, we're a lazy bunch around here after school's out. When Herb's home, he gets us up."

Then Dexter, a boy my age, came down. He was a gangly kid, quite a bit taller than I was. We'd been introduced the night before.

"Hi," he said.

"Hi."

Finally Lorna, who was going to nursing school, came down. She was very pretty, I'd noticed, but besides being four or five years older than I, we were related. So I squelched my attraction to her.

At breakfast, Dexter said, "Wanna ride around town with me?"

"Sure."

"You ride a bike? You can use Lorna's, can't he, Lorna? It's a girl's, though. That OK?"

"Sure."

After breakfast we got on the bikes and went riding down the shady streets. Some blocks away we picked up another kid, then continued riding. It was a nice day, and Lima seemed like a nice town. It was a city, actually, of about thirty thousand, Dex told me. We passed two girls, walking along the sidewalk.

"Hey, Dex," the other guy said. "You see the knockers on the one?"

"Like headlights," Dex agreed.

He and his friend started up a joking, tough-guy banter. I was Dexter's hick cousin, being shown the fast life in the city.

We were passing a neighborhood grocery store when Dex said to me, "Hungry?" He and his friend exchanged looks.

"Sure. But I don't have any money."

Dex grinned. "You won't need any."

We swung our bikes around and dismounted in front of the store.

"Just follow us," Dexter said out of the side of his mouth, "and don't say anything."

We walked into the store. The grocer looked up at us, then turned back to a customer.

Hidden among the stacks, Dex and his friend started grabbing goodies and stuffing them into their shirts. My mouth must have opened, but I kept quiet.

Outside and riding unhurriedly away on our bikes, Dex said, "Whattaya got?" His friend produced a handful of treats. Dex had another handful.

"Who's ready for pop?" he asked.

We rode into a gas station, paid the attendant some nickels, and fished bottles of Pepsi or Coke, root beer or orange crush, out of the ice water in the cooler. I ate my share of the stolen treats and washed it down with a Coke. My cousin and his friend continued their smartass bandinage while I was seized, suddenly, by euphoria. "Jeeze, how come this stuff tastes so good!"

Dex winked at his friend. I was a hick all right.

When Dex and I got back to the Wilson house, we were promptly sent out again by his mother—to the store where we'd just shoplifted—for a few groceries. In the store we walked past the unsuspecting owner, who smiled at us from his cash register, and picked up a bottle of pasturized milk, a loaf of white bread, and a pound of "ole." The oleo-margarine, white with a yellow dot in it, looked like a block of lard, but outside the store Dexter began to push and squeeze at the colorless gunk, spreading the dye concentrated in the dot until it turned the whole block into a buttery yellow. "The low-price spread," Dex called it.

There was something so city about that, I thought in my country snobbishness. Oleo-margarine wasn't butter and it didn't taste right on toast in the morning, nor did pasturized milk taste right—it tasted "off" to me, thin and sour.

A couple of days later we were heading home.

We avoided Chicago this time, and instead drove straight across Indiana and Illinois to Davenport, Iowa, then followed the Mississippi up to the Twin Cities. Again, we must have stayed overnight somewhere in a motel. The next day we crossed the Mendota Bridge and drove through Minneapolis and were practically home. From downtown Hennepin Avenue, our farm was less than thirty miles away.

When I went to Confession soon after our return, I didn't mention the shoplifting incident in Lima, though I felt somewhat guilty about it. But had I actually taken anything? No, I decided.

I gave myself absolution.

A day or two after we got back from Lima, my father pulled our new baler out of the shed and greased her up. He'd called around and got some custom work. We had hay of our own to bale and corn to cultivate, so the work would have to be staggered, he told me. He also said I could forget about playing this summer. I'd be working.

We set out after breakfast that morning to bale hay for a farmer north of Hamel. It was late June, the start of a hot day, good haying weather. I stood on the drawbar of our 1939 International F-20 Farmall, a big old tractor we'd acquired after the war, balancing to its rocking motion, one hand loosely holding the back of the tractor seat on which my father, a hand loosely on the wheel, sat driving us down the gravel road. Towed quietly behind us on its rubber tires was the expensive machine my father had bought and had to pay for now, a New Holland Model No. 76 automatic baler, gleaming red and yellow in its factory paint, with the blue Wisconsin engine that drove it as yet unmarked by workaday scratches, grease stains, exhaust soot. Whining along at ten miles per hour with the Farmall in road gear, my dad and I grinned at each other.

By late afternoon, we weren't grinning anymore. I was on the tractor, craning my head around for the signal to start up again, and my father was on the baler seat, struggling to adjust the damn thing. He'd had the manual out repeatedly, had read and reread the directions. Still, nothing worked right. The bales came out too long or too short, or one or both of the knotters failed to tie so the bales burst apart as they left the chute. We were stopped in the field more than we were moving, and time was wasting. My father's temper was getting short. Then the farmer whose field we were trying to bale came out to offer useless advice. I waited for my father to explode, but he could hold off when he had to. Once the farmer was out of hearing, though, he let go with every foul-mouthed expression he'd learned not on the farm but as a tough kid on the streets of Minneapolis. His swearing seemed to flatten the landscape; you felt the shock waves. "What the hell's *wrong* with this fucking thing?" he asked me furiously, as if I should know. "What's a man to *do*, for chrissake?"

Then he pressed the metal tab that shorted the spark and shut off the baler motor. Got out the manual again. Took a deep breath and forced the air out through pursed lips. Studied the diagrams. Reread, for the umpteenth time, the instructions. Readjusted the bale length (a little over three feet) and set the knotters again, step by careful step. Then he yanked the starter rope and got the Wisconsin going again and put the baler in gear. Nodded to me on the tractor. I shifted, advanced the throttle a little, let the clutch out. We started down a windrow, the rotating pickup guiding the hay up the moving apron into the piston-like baling chamber. I looked back. My father sat poised on the baler seat, watching the knotters. He looked up at me, nodded; looked back at the knotters. *Ka-whump.*

Ka-whump. The plunger lifted and fell, pushing loose hay into the bale chamber to be sliced and compressed as it was forced into the chute. At the correct interval, twin needles plunged through the packed hay and two strands of sisal twine wrapped around a section, the knotters worked, the counter advanced, and a perfect bale of the right rectangular size and tightly bound by the twine dropped like a huge loaf of coarse bread onto the field.

I glanced ahead to keep the baler on the windrow, then looked back to see my father's big grin. He twirled an arm over his head, meaning crank her up! I advanced the throttle, aimed the tractor down the field. *Ka-whump. Ka-whump.* The Wisconsin roared. The heat from the tractor engine blew in my face. We were baling now. I drove around and around the field, and began to doze from the heat and the noise and the monotony. "Wake up!" my father yelled. That's how it went through the afternoon.

And that's how it went through the summer. It was a summer of abrupt change for me. All at once and all day, I was in the fields with my father or sent out there by myself to do a man's work, so I had to keep my mind on it and *pay attention, goddamnit!*

That was the summer I felt, nearly every day, my father's frustration and anger, his anger at me, for this or that fuckup, for daydreaming, for not paying attention.

We were baling one day, our own or a neighbor's hay. The knotters were acting up again. I was on the tractor, stopping and starting at his angry signals from the baler seat, getting a stiff neck from cranking my head around. We had stopped and I was daydreaming. Looking back finally, I caught my father's eye and he raised and lowered his arm. I thought that meant we could start up again, and I advanced the throttle, put the tractor in gear, and got it moving. Je-zus Christ! I didn't actually hear the words above the sound of the machinery but read his contorted lips as he jumped from the baler and swam through the air toward me. I stopped the tractor and waited for him.

"How many times do I hafta tell you?" he shouted. "When I go like *this*"—his right arm swung up, then down—"it means stop! When I go like *this*"—identical gesture—"it means go!"

That became one of our funny stories about the farm, one I'd tell at family gatherings and receive the relatives' uproarious laughter and my father's weak protests. "Aw, I ain't that bad, am I?"

His rages *were* kind of funny—when they weren't frightening, or when his words weren't belittling or humiliating, telling me I was no good and would never amount to anything. When I wasn't so scared of him anymore and knew his bark was worse than his bite (though his bark could be pretty scary)—knew how to retreat into myself and hang on during the storm of his punishing words—he reminded me of Yosemite Sam, that comic little figure of rage in movie cartoons. Red faced, fuming, you could almost see smoke coming out of my father's ears. You wanted to laugh, but you knew better.

※

That was the summer I continued helplessly to live in the imaginary world I'd created within the real world of the farm. It was Kianga's world. Kianga: my wild boy self, who I'd played in the woods of Elder's Forty or on the marshy and wooded shores of Spurzem's Lake, until I had no time anymore for play; and besides, I grew a little too self-conscious to act him out. Then he went underground, became my secret self represented by the small (like me) snub-nosed pliers I always carried with me now in a back pocket of my jeans as another boy or a girl, clinging to childhood, might lug around a tin soldier or a doll.

I made Kianga's world by the simple device of imagining inches were feet and thus multiplying everything by twelve. So my five-inch pliers became five-foot Kianga. So a mile became twelve miles, a forty-foot tree stood four hundred and eighty feet high, and the quarter-by-half-mile dimensions of our eighty-acre farm enlarged to a three-by-six-mile medieval estate or ancient little kingdom.

And so the real, ordinary world I inhabited was magnified into a colossal, heroic world in which Kianga lived and had his adventures.

He filled my inner life. My father could yell at me to *pay attention. Take an interest, for chrissake,* but he couldn't drive Kianga out of my head. Especially when I was alone, out in the fields or in the barn, past the lost, playful days of my childhood and into the workaday seasons of my adolescence, I lived in and through him, kept up a running story about him . . . until that day on the farm when I was shocked, once and for all, into reality.

※

But of course the real world of the farm was always there, and I had to live in it. I had to, after all, pay attention, take an interest. And, yes, hefting bales of hay, lifting full, ten-gallon milk cans over the rim of the cooling tank and lowering them into the water, shoveling shit from the gutters, I felt my growing strength. Looking in a mirror I saw the enlarged muscles in my arms, their popping veins. I was a little bastard, smaller than almost every other boy around me, those strapping Germans and Scandinavians, even most of the Frenchies, but I was wiry, my upper body tanned dark from going shirtless all summer. I had "Indian muscles," my mother told me one day. She told me what I wanted to hear.

We were custom baling one day, north of Hamel. Toward evening, my father thought he might finish the job if he could keep going. He decided to send me home for supper and to do the chores. Even without cows to milk, there were our fifty or so chickens to feed and their eggs to pick, our calves to feed and be bedded down. Our heifers were in pasture, but they might have gotten out and into a corn field that would bloat and kill them. I was to eat, then bring Dad some food back and relieve him on the tractor so he could eat.

    We took our truck with us, our practice on a job now—we could use it to run for parts, or pop, or to the restaurant outside Hamel for the noon meal, though my mother usually packed us a lunch of sandwiches and jars of iced milk. "Take the truck," Dad said. "Watch for cops."

The boy on the place, who was my age and, like me, too young to drive legally, went back with me. I could see he was impressed by my father's letting me take our truck over public roads.

While my mother fixed supper, we played catch in the yard with a tennis ball. Like me, this boy would start high school that fall, though at Wayzata, not Mound, where I would go. We had other things in common: we were both farm kids, both growing into young men. His easy acceptance of me was something new and pleasant. Suddenly I felt part of an age group, our age group, and no longer was I just a solitary farm kid lost in his fantasies.

After supper I drove us back to his place, where my father had almost finished the baling and this boy's father and an older brother had started loading the bales onto a wagon behind their tractor. The boy's sister was there with lemonade for all of us. She was way too young, about twelve, I guessed, but very pretty and developing. I was beginning to notice all the pretty and developing girls.

Toward the end of that summer my father bought a purebred Holstein bull and a dozen cows, and we were back in the milk business.

Then a week or two before school started, my face broke out in little red pimples. Though I scrubbed them with Lava soap till my skin was raw, they stayed. It was about this time, too, that I woke one night thinking I'd wet my bed like a child. Then I realized what it was; I'd heard about it. After that I tried to program my body before going to sleep, tried to will it to happen, but I never could. Just as well, because causing it to happen might make it a sin of the flesh—a mortal sin—which I would have to confess. Wet dreams, though, were involuntary and therefore innocent—weren't they? I decided they must be.

A HICK IN HIGH SCHOOL

Five of us Klatte kids got on the school bus that first morning after Labor Day in 1949. Since the year before, we'd been able to ride the school bus and be dropped off for grade school in Loretto because by then there were kids on our road attending the public high school in Mound and the bus route went through Loretto. I wouldn't get off there this morning. I was starting high school.

In Loretto, my sisters and the twins got off the bus and headed up the street to SS. Peter & Paul. The twins, turned six, were starting first grade. Marcia, age ten, was in the fifth grade. Joyce, thirteen, was in the eighth grade that year and eager to get it over with so she could go to high school with me.

The bus pulled out of Loretto and headed south, past Lake Independence, to Maple Plain, where a dozen or so kids got on; then on through the rolling countryside, past woods and fields and clusters of farm buildings, to Mound, on Lake Minnetonka, and the high school on its main street. The main street, including the town's business section, wasn't strictly on the lake, but its residential sections were, so that Mound was as much a lake community as Wayzata, at the other end of the lake, or any of the other communities along Minnetonka's many bays and inlets. It was some twenty miles by road from our place, and the bus trip, with its many stops, took more than an hour.

The school collected farm kids from the district, as well as town kids, lake kids, whose parents made their living in or around Mound, or whose fathers went to work in Minneapolis wearing a suit and tie. Mound wasn't as rich as Wayzata, but you had to be white and somewhere in the middle class in those days to live anywhere around Lake Minnetonka with its woods and water and lovely little tree-lined roads.

That first day in Home Room I was amazed by the raucousness, the undisciplined horseplay of the students. Sister Superior would have had those kids sitting quietly in their seats and copying things off the board. Gradually, as I sat grinning and looking around the room, acting as though I was used to such behavior, I realized that if for me the ninth grade was my first year of high school, for this rowdy bunch—mostly boys, it seemed—it was their last year of junior high; they'd already had a couple of years here. They knew the ropes, knew what you could get away with.

In class I couldn't break myself of the habit, learned at sister school, of standing up when the teacher called me. I'd hear titters behind me. Once I heard a girl whisper, "How come he always stands up like that?" and her friend whispered back, "That's what they do, those hicks." Pretty soon I was slouching in my seat and cracking wise, or trying to, just like the other kids. I didn't want to be a hick.

Yet in the halls of the school the town girls crinkled their cute little noses and said, "Pee-you!" or "*Farmer!*" as I passed.

This even though I'd washed and changed my underwear after cleaning the barn that morning, wore a clean shirt and the requisite jeans, their cuffs turned back in a couple of tight, stylish rolls (not, as on my first day, in one big, farmerish roll), had my hair slicked back and scented with Brylcream or Vitalis and my armpits doused with Mennen's spray deodorant. The barn smell never left me. It was in my pores. I might as well have gone to school wearing bib overalls and a feed store hat and chewing on a stem of hay.

On the school bus, though, nobody seemed to notice my barn smell or anybody else's, probably because the bus was mostly filled with farm kids. I sat in back with the other boys, where the sway of the bus and the smell of exhaust made me sick. After the long, enervating ride to or from school, I'd feel dizzy and nauseated.

Nick Herman, already a sophomore at Mound, was on bus patrol that year when Hal and Dan and I wouldn't have been caught dead wearing his silly arm band. You could practically see the swastika on it. The patrolman kept order for the driver and jumped off the bus at railroad stops to flag it across. The beaming pleasure on Nick's face as he climbed into the bus again and took his seat behind the driver prompted us to yell "Heil Hitler!" from the back of the bus. Nick just grinned and gave us the finger.

I didn't have a driver's license yet, but I'd been driving our farm truck on public roads for a year or more and now the folks let me take their car on occasion. At first it was just to drive to Loretto for religious instructions. Father Michael, hoping to bolster our faith against the bad moral influence of the public high school, held instructions for ninth-graders, us escapees from SS. Peter & Paul, every Wednesday night in the school. It was your religious duty to attend. I'd start chores with my father, then have to remind him it was instructions night.

"Oh, all right then. You better get going."

"Okay to take the car?"

"Yeah, yeah. Just be careful."

I'd wash up in the house and change into my school jeans and loafers, put on a clean shirt, then climb into the folks' car, a '46 Ford, and start her up. It didn't matter that I was going to religious instructions. This was my night off!

Father Michael's instructions were pretty dull (they, like SS. Peter & Paul itself, seemed like something I'd outgrown), but his rants about high school being our damnation unless we kept our faith, together with discussions of such things as dating and petting—sex, in a word—sometimes perked the evening.

Afterwards, driving home, I'd be tempted to tromp on the gas, see what the old car could do, but my father's certain wrath, should I have an accident, kept me well under the speed limit.

Eventually I got the nerve to ask for the car to go to a show. Dumping milk for my father during evening chores, I'd gear myself up to it. Finally: "Daddy, can I have the car tonight to go to the show in Wayzata?"

He never turned me down. What he usually said was, "If it's all right with your ma, it's all right with me."

What Ma usually said was, "If it's all right with your dad, it's all right with me."

So I got to drive to Wayzata whenever there was a Randolph Scott or Joel McCrae western, or a monster or science fiction movie. Hal and Dan and Cal Tyson went along with me, and I suppose it says something about country living then that a fourteen-year-old kid without a driver's license could get the use of the family car and, moreover, take his buddies along as passengers. That was a more innocent, or anyway more trusting, time, and the cops weren't as watchful either.

The other place I could drive to was Independence Beach, a dancehall on the lake where there was roller skating Friday nights. Not having grown up around sidewalks, I'd never roller skated until now, but I could ice skate a little—straight on, no fancy backward stuff—so that skating on those little wheels you clamped on your shoes came easily enough. To the sprightly, awful accompaniment of recorded organ music, we'd roll thunderously around the wooden floor, round and round, raising dust that stuffed your nose and made your eyes water.

But there were girls to skate with. Holding both hands with a girl in an awkward, crossed-armed way, you joined the flow of traffic around the floor. It was more uncomfortable than romantic, the excitement coming when somebody fell and there was a pileup of arms, legs, and metal skates—screams and laughter, blood sometimes. Helen, though, was different. She was in the eighth grade with my sister Joyce in Loretto, a friend of Joyce's, and my secret heartthrob. I'd gulp back my abject fear of rejection and ask her to skate. Helen's hands weren't rough and calloused, I noticed, like most farm girls' were. That meant that, like my sisters, she wasn't made to do outside work. She smelled of clean sweat, and her face glowed. By the end of the night I was so choked with allergies and infatuation, I was truly miserable.

※

I had dropped off Dan and Hal after roller skating one night, and was already home when I realized I'd left my sweater at the rink. It was my best wool sweater, a Christmas present from my grandma Loomis, and I'd taken it off at the rink because I was hot and left it in a booth. It was midnight already. The place was probably closed by now, but anyway I jumped back in the car and started driving, too fast, toward Lake Independence.

Where the road became a narrow causeway through the marsh by Spurzem's Lake, I met a car. Its lights blinded me, and I hugged the edge of the road to miss it. *Clump, whump.* My car stopped, hung up on the bank.

I gunned the engine a few times, shifting from low to reverse, trying to rock it free, but the car only tilted more precariously toward the cattails. I shut off the engine. Got out under the stars and tried to think what to do. Decided I needed help.

I walked to the nearest farm, the Millers', and knocked on the house door. Knocked again. An upstairs window opened and Willie, who had worked as our hired man when not much older than I was that night, poked his head out.

"Who's there?"

"It's Ross. Ross Klatte," I said. "I'm stuck, down by the bridge."

"Just a minute."

The window closed and, presently, Willie came out. He was grinning.

"Stuck, huh? Your folks know you got the car?"

"Not exactly."

Willie winked. "This'll be our little secret." I followed him to the machine shed, where he started up their John Deere and backed it out. "Jump on."

I climbed onto the tractor's drawbar and we putt-putted down the road. In the tractor's lights, my folks' car looked ready to tip over into the swamp.

"Oh-oh," Willie said.

"You think we can get it out?"

"We'll see."

Willie chained the car to the tractor and took up the slack. "Get in and put her in neutral," he told me.

Then he throttled up the John Deere's two-cycle engine and eased the hand clutch out. Holding my breath and steering toward the tractor, I felt the car lift easily off the bank and settle onto the road.

I got out and said, "Jeeze, thanks a lot, Willie. How much do I owe you?"

"How much you got?"

I reached for my billfold.

"Put it away. I can't take your money. Drive careful."

Then he putt-putted away into the night.

I never got my sweater back but years later, when I told my father about running into the ditch that night, he just smiled and shook his head. "Hell. That's the first *I've* heard of it."

Good old Willie Miller, prematurely dead by then, had kept our little secret.

There was a roadhouse outside Loretto, on Highway 55, that had closed recently and been locked up. It seemed to have been abandoned. One night after religious instructions I proposed a "raid" on the place. I'd drive by slowly with the car's lights out and Hal and Dan would bail out. Then I'd cross 55 and pull over. Wait a while. Then turn around and drive back, lights out again, to pick them up. It would be like dropping commandos by parachute behind enemy lines—or like something Tom Sawyer might have dreamed up.

Hal and Dan came back with some bottles of beer.

"There's cases of it!" Dan said. "Just there for the taking!"

The next week, after instructions, we broke into the place again. Only this time it was my turn to take the risk, Hal said. "I can drive your old man's car. You and Dan go in. Next week Dan gets to drive."

That sounded fair. My heart was thumping when Dan and I jumped into the dark from the moving car and rolled into the ditch. I sat up and watched the car lights flick on after Hal had crossed 55.

"C'mon," Dan said. "I'll show you where we get in."

He led me to the back of the place and a small window he pushed open. "It's the window in the women's can. They forgot to lock it."

We climbed onto the outside trim and then into the building. Felt our way into the bar. "Let's see if they left any hard stuff," said Dan.

Somebody else had been doing mischief here. The floor crackled with glass from broken windows. Suddenly lights flicked across the walls.

"Cops!" said Dan. We flattened ourselves against the glass-strewn floor.

A highway patrol car had driven in, its headlights crossing the dark roadhouse, and stopped by the entrance. We could hear its motor idling. Then their spotlight came on and shone into the place. Just like in a movie Dan and I lay still on the floor, suspending panic, as the cops' spotlight played above us. Finally they drove off.

"Jesus," Dan said.

"Let's go!"

"Wait," he said. "We can't leave empty handed."

We went to the back room, our eyes somewhat accustomed to the dark now, and helped ourselves to a case of Grain Belt. It was a job getting it through the window in the women's can, but we did it, then stumbled through the dark to the road.

Hal came by in my folks' car—it was the second time, he said—and we piled in. "God!" he said. "I saw the cops and here I was driving with my lights out. And driving somebody else's car without a license. They woulda thrown the book at me!"

"They woulda thrown the book at all of us!" said Dan.

Hal slid over and I got behind the wheel. I drove us to a dark side street in Loretto and we divided the beer. At home, I hid my eight bottles behind the outhouse and the next night, after everybody else was in bed, I sneaked out and forced myself to drink some of the beer. The fact was, I didn't like beer then. My preference, when I could get it, was for sweet Mogen David wine. So what I did, after another night of surreptitious drinking, was sneak the remaining bottles out to the rock pile across our swamp pasture one day, stand them up, and "pop" them with my .22 rifle.

We decided not to push our luck by staging another break-in. If we were caught, we'd be charged as juvenile delinquents, we thought, and maybe even serve time in the Red Wing reform school. There was something faintly romantic about the idea, but none of us wanted to chance the reality.

Dan went to confession about it.

"You didn't mention our names, did you?"

"Father didn't ask," Dan said. "But he said I had to make restitution, replace the beer or pay for it. I had to promise before I got absolution. You think I can just put some money in an envelope and mail it to the place?"

"Maybe."

"We shouldn't a done it."

"You shouldn't of confessed it!"

"Didn't you?"

"No," I said.

The place reopened eventually, under new management, and the next time we saw the inside of it was one night after our graduation from high school. On the way to a dance in Waverly, we had run off the road. Hal, who was driving the car he'd just bought, nearly broke his nose against the steering wheel and I hit the windshield, leaving a spider-web pattern on the glass and raising an egg-size bump on my forehead. Dan wasn't injured.

We struggled up out of the ditch to the road and were picked up by somebody who took us to the roadhouse we'd once broken into. There the waitress, a good-looking Loretto girl who'd been three or four years ahead of us in school, made a pack of ice from the bar, applied it to my forehead, and gave me a thrill.

Not all of the girls at school smoked and wore tight jeans. Most, actually, wore the ankle-length skirts then in vogue and plain blouses. But there was one girl who epitomized what I'd expected, and I longed to know her. She was small, dark skinned and slender, with long, slightly curly black hair, an exotic contrast to the blond looks of so many of the other girls. She had a nice little ass under her tight jeans and pointy little breasts under a succession of frilly blouses. Her mouth was a slash of ruby lipstick, and she spoke out of the side of it like a gun moll. She was pretty in her tough way and yet fragile looking too, like Ida Lupino in the movies, and I stole glances at her in the couple of classes we shared. I met her appraising eyes once and the blood pounded in my head. I thought she was maybe part Indian or a gypsy. "Naw," some town kid informed me. "She's a little Jew! Wouldn't you like to just bang the shit out of her, though?"

The boys' locker room was an education in itself for a kid raised by a devout Catholic mother and schooled by nuns. Testosterone hung like musk in the air along with steam from the showers and the rank smell of unwashed sweat clothes and filthy jockstraps. There was talk of "making out," of "jacking off," cloudy, disturbing terms to me. One guy, handsome, cool, older—he'd flunked a couple of grades—was a known makeout artist. He had a one-syllable name; I'll call him Burke, and the legend "Burke eats box lunch" often appeared on the coach's blackboard. I wondered at it until somebody explained to me that "box" was another word for pussy. Then I still had to wonder.

Burke had a follower, neither so handsome as he was nor with any of his style, who showed off the rubbers he carried in his billfold and got Burke's "leavings," according to locker room gossip. One day, as if to prove to us that Burke had nothing on him, he lifted his shirt and there, on his back, were the telltale scratches. That was the badge, the proof. He'd actually done it with a girl and in her excitement she'd left those enviable marks on him. Then, to further impress us, he pulled a snapshot out of his billfold and passed it around. It was a closeup of a naked, reclining woman (her head didn't show in the picture) with splayed breasts and a hairy cleft between her legs. "You ever seen *that*, you farmers?" the guy said. "Ever seen any cunt except on a cow?"

Like most farm boys, I went out for wrestling. It was our sport—that and football, if you were big enough; track, maybe, in the spring.

Basketball, "pussyball," was for town kids. It vied with football as the school's most popular sport, and its players attracted the most desirable girls, the "queens." Among the queens the most lithesome, the most desirable, were the cheerleaders, whose bouncing, arching, gorgeous leg-flashing displays during a game truly roused us on the bleachers. Basketball itself, I had to admit, was exciting—a dazzling, non-contact ballet of a sport in which speed, teamwork, skilful ball handling and spectacular swishes brought more or less painless glory.

Wrestling, by contrast, was hand-to-hand combat, more strenuous even than farmwork, and more often than not pitted you against another muscled farm kid.

It was about the only sport I could sign up for because it was divided into weight classes, so that even a runt like me could compete. I wrestled in the lightest class, as a ninety-five pounder, but found myself struggling at practice against guys who, unless they starved themselves for a match, weighed something more than ninety-five. Myself, I weighed eighty-nine pounds that year.

Sports practice was the last two hours of the school day. You skipped study hall and instead went down to the locker room after your last class and changed into jockstrap, sweat clothes, and sneakers. Then you went upstairs to the combination gym and auditorium, where mats had been spread in the pit between the spectators' seats and the raised gym floor. We warmed up with a few stretches, some pushups, situps, then lay back and waited for the coach to pair us off while sneering up at the willowy town boys running back and forth on the gym floor, dribbling, shooting baskets, having fun. We, on the other hand, were gladiators, waiting to enter the arena.

"Takedowns!" Coach would order.

We'd pair up and take our turns at charging out from the edge of the mat to circle, feint, then grapple with each other for the takedown, and the first two points.

"Break!" the coach said. "Next!"

Then it was time to practice the referee's position. In this, which you and your opponent assumed at the start of the second and third periods or after going off the mat during a match, you flipped a coin for the choice of "up" or "down." If down, you went to your hands and knees and the other guy knelt behind you, wrapped one arm around your belly, and grasped your upper arm with his other hand. Then at the command, "Wrestle!" you tried to get out from under the guy. The classic manoeuver was the "sit-out." You threw your legs out, arched and twisted, and broke away. One point.

Or, and this took more strength and skill—you threw your legs out, twisted, caught the inside of the other man's leg for leverage, then pulled yourself over onto his back, reversing your position. Two points. Or you simply pressed your arm against his arm around your middle and rolled, trying to wind up on your back on top of him and, ideally, after twisting around, achieve a pin hold. The object was to pin your opponent, pressing his shoulders to the mat for the eternal, two-second count. To achieve this took speed, agility, strength, endurance, and a couple of other things. "You gotta have heart," the coach told us. "You gotta be fast—both above and below your ears."

We were issued a booklet on Greco-Roman wrestling that included descriptions and illustrations of the many holds. "Read it," coach said. "Memorize the holds. Try 'em out at practice."

A day or two before a meet, the coach would announce, "Eliminations!"

That meant matches between the best boys in each weight class to determine who would wrestle on the team.

It was exciting, scary, though bloodless except for the occasional nosebleed, from exertion or from an accidental blow. Practice was more or less play, but an elimination match was as earnest and real as any match against a rival from another school.

But I was second string, and so eliminations caused only a little initial queasiness. Once I'd been passed over, as usually happened that year, as not good enough to wrestle on the team, I could sit off the mat with the other second stringers and just watch the eliminations.

One Saturday in April Dad and I were sacking up corn and oats to take to the feed mill in Hamel when spring started—started, as it often did in Minnesota, all at once. We'd filled maybe a dozen gunny sacks with oats from the granary and loaded them onto the truck, and we were in the corncrib now, filling sacks with ears of corn. After we'd loaded those onto the truck, I'd drive to the mill and have the corn and oats ground together for cattle feed. I had my license now (after flunking the driving test the first time, despite the year or so I'd been driving illegally; I had to learn how to make a proper left turn). I was being trusted with the job of going to the feed mill by myself.

It was a warm day, with the sun flashing off the snow in the fields beyond the farmyard. I was holding the sacks and Dad was doing the shoveling. For a while I'd been dimly aware of a new sound in the air, coming from below the barn. Then I knew what it was.

"Listen," I said.

Dad paused with a shovelful of corn in his hands.

"The crick's running!"

He grinned. "Go and see," he told me. "I'll use the sack holder for a while."

I ran down through the snow to the little creek below the barn. That morning it had been low between its banks and iced over. Now it was full of gurgling water, rushing over its frozen bottom to the slough that was our swamp pasture in the summer. The snow was collapsing along its banks, sliding into the water. There was water at the head of the slough and a growing dark area of slush beyond it.

"Whooee!" I cried. I'd stepped too close to the creek and my overshoes filled with cold water. I took them off and, standing in the snow in my stocking feet, poured the water out of them. "C'mon," my father said, when I got back to him. "Let's finish this job so you can get to the mill."

That evening the creek slowed, then stopped, as the temperature dropped back to freezing and ice formed over the slough.

But the next day was warm and sunny again, and the creek started flowing again, and the slough filled with water. By the following Saturday, it was a pond for sailing toy boats across.

Then the ice broke up on the lakes and the birds were back and the trees began leafing out in the sunshine. And I had spring fever.

The tramp

Spring fever. It was a yearning to somehow lose myself in the greening world, to be free as the birds, free as a tramp like Gearhardt Moe.

 Of all the hired men we'd had—including toothless old Pete Gears, who could chew steak with his gums; the retired Great Lakes sailor, who danced Celtic jigs for my sisters; Jack, the shellshocked war veteran; the alcoholic young man who showed up at our place one morning after waking up in a ditch and, though terribly out of shape, amused my father and me after he'd gone to get the cows by clearing the pasture fence like a track star when the bull was after him—Gearhardt was the most memorable, the strangest, the freest. Twice he entered the world I was growing up in, that was starting to imprison me, and twice he left it when spring came. When I was fifteen years old, I wanted to be like him. Like him, that is, without his smell.

He came to us the first time one fall evening after the war. It was in October, I think, when frost formed as soon as the sun went down and we were still pasturing the cows during the day but keeping them in the barn at night. I was dumping milk for my dad, carrying a pail out of the barn to the milk house, when a man appeared in the doorway. He was tall and dark, in a slouch hat and ratty coveralls. His hawk nose poked out of a beardless face, and his black, greasy-looking hair hung to his shoulders.

"The man about the place?" he asked in a soft, quick voice.

"Back there," I said, pointing down the line of cows.

He brushed by me, almost knocking me flat with his smell. It was the smell of a man who hadn't bathed in months, years maybe, a fetid, yeasty, musky soursweetness born of all the body's unwashed secretions. Most of the hired men we'd had stank a little like that just coming off the road, but Gearhardt's smell was special. It was beyond unpleasant. It was a wild smell. It was how people must have smelled in the days when bathing was considered unhealthy, even sinful. It was a force of nature.

When I stepped back into the barn from the milk house, swinging the empty pail, he was standing in the walkway where my father had just moved from between the cows with the support belt for a Surge milker over his left shoulder and the full bucket in his right hand.

"Yeah, we got plenty of work," my father was saying as he dumped the milker into a pail. I picked it up to take to the milk house. "I'm working out this winter and my boy's in school, so I could use a man that knows cows. You eat anything today?"

"Nope," the man said.

Dad turned to me. "Dump that milk, then take this man up to the house and tell your ma to fix him something."

"No-no," said the man. "I can wait till you finish up."

"Okay. You can throw hay down then. This is Gearhardt," my dad told me. Then to the new man: "Go with Ross here. He'll take you to the mow."

Many barns had a ladder to the mow, but with ours you walked outside of the cow barn and up and around to the big sliding doors that opened into the vastness of the hay barn. There was a regular door built into one of the big sliding ones and a switch just inside that turned on the light above the hay chute. This was before we had a baler, and the two mows on either side of the floor between them were stacked to the crossbeams with loose hay. Dad had forked some out during the day and piled it by the chute. You could look down through the chute to the manger and see the cows in their stanchions. You could feel the rise of the warm, moist air from below, that, in the winter, would plaster the beams above the chute with hoarfrost.

Gearhardt nodded and started forking the hay down. I left him and walked down to the cow barn to spread it. Dad was finishing the milking. The new man came by as I was bedding down the cows, forking fresh straw under them.

"Here, boy. I'll do that now."

He took the fork from me and got more straw from the bin. He spread it deftly beneath the cows, cooing "So, boss, sooo." Some animals swung their heads in their stanchions to watch him; others stopped chewing, just a little nervous. But none kicked at his fork or even shifted her stance.

"Look at that," said Dad. "He's a stranger to'm, but the cows accept him. Now *there's* a man been around cattle."

I was both relieved and resentful towards him for taking over some of my work.

In the house after the milking my mother's face stiffened and her nose crinkled a little, but she warmed up leftovers from supper and we watched Gearhardt eat. He ate hungrily but rather daintily, with an air almost of refinement.

"What's your name again?" asked Ma.

"Gearhardt Moe," he said quietly. He jerked when he talked.

"What nationality is that?" Dad asked. He was always interested in a person's origins.

"German."

"Thought so," said Ma.

"I got some Polack in me too."

After his meal, my mother hemmed a little and said, "Would you like to take a bath, Gearhardt? I'll heat some water for you."

"No-no, Missus."

She glanced at my father for support, but he didn't say anything.

"Then I'll show you to the hired man's room." She started for the stairs.

"No-no. I'll sleep in the barn."

"Naw," my father said. "No hired man of mine has to sleep in the barn."

"Maybe tonight would be all right," Ma said, looking pointedly at my father. "I can give him blankets and a pillow."

"That all right, Gearhardt?" Dad smiled ruefully. He'd finally gotten the message from my mother.

After Gearhardt had left for the barn, Ma said, "He wasn't *always* a tramp. He stinks to high heaven, but he has manners."

⁂

Gearhardt clanked when he walked. That was from the collection of little bottles he carried in the pockets of his coveralls, we noticed. They were an experiment, he said. What *kind* of experiment? He was working on a patent medicine. "Gonna make me rich." (I found a cache of his little bottles under a fence post in one of our fields one day, after he was gone that first time. I screwed the cap off one and smelled it. Just as I'd thought: stale piss.)

Gearhardt slept in the hay barn for a couple of nights. Then, at my mother's insistence and with my father's backing, he took a bath—*His water turned black*—and got a haircut in Hamel. Ma washed and mended his torn, filthy coveralls, his rotten socks. *You could have stood them up against the wall.* Washed and with his hair cut, dressed in clean and mended coveralls, Gearhardt was transformed. He wasn't bad looking until he smiled; then you saw his mossy or missing teeth. His age was anybody's guess, but he wasn't old, the folks thought—in his thirties maybe.

He was a good man, good with cows, as my father said, and in fact spent most of his time down in the barn with them.

"I think he likes cows better than women," suggested my father to my mother. She made a face. The implication went over my head.

He couldn't drive or operate any of our machinery, including our Surge milkers, but he could clean the barn, he could curry the cows, scrape the "dingle berries," the balls of dried shit, off their backsides, and he could feed and bed them down. So long as he was with us my father had a man on the place to look after the cows while I was in school and he was working out during the winter. So long as we had Gearhardt I got out of cleaning the barn on school mornings and had some free time on weekends.

He told us a little about his life on the road. Slept in barns and haystacks. Only worked when he had to, otherwise roamed the country, "looking." For what? we wondered. "Nothin'. Just seein' the country."

He didn't drink, didn't smoke; he'd seen what drink could do to a man, and smoking made you cough.

Didn't panhandle. Wasn't a grifter, pretending to be blind or crippled; didn't roll drunks.

Didn't go south. "Too many 'bo's down there. Jack rollers. Too crowded."

In the old days before the war he'd found work in the fall harvesting wheat in the Dakotas and in the winter cutting ice off the lakes around the Twin Cities. But there wasn't much threshing anymore—they used combines on the prairies now; eventually we'd have one—and they didn't cut ice anymore either because people had freezers and refrigerators now.

Rode the freights when he had to. Walked, mostly. Didn't hitchhike (who would have picked him up?). Took his summers off, sticking to the country, living on bread and peanut butter, canned stews or baked beans heated over a fire or eaten cold; corn stolen from the fields or vegetables from people's gardens, helping himself after dark. Slept in hay barns or camped in the woods by country roads. Avoided the cities, "main stems," hobo jungles.

You know, you can go a long way on just a candy bar.

He stayed with us through that winter. When spring came, he walked away.

❧

It was a couple of winters later, I think; I was down in the barn after having cleaned it and was feeding the cows silage when I heard our car drive into the yard and then my father's rap on the barn door. He was late, and I'd been thinking I might have to skip supper for now and start the milking. I unlatched both halves of the Dutch door and my father swept in through the fog of barn moisture escaping into the cold.

"I ate in town," he told me. "Go on up to supper. I'll start the milking."

He was grinning for some reason.

I walked up to the house. On the porch I stomped the snow from my overshoes, then unbuckled them and pried them off with my toes and left them by the door. I stepped inside to that familiar, overpowering stench.

He was sitting at the kitchen table, his back to me, but turned around just as my mother spoke from the stove.

"Gearhardt's back!" You could hear the false enthusiasm in her voice.

"How-do, boy?" he said in his soft, jerky way.

He was in the same condition as when he'd come to us the last time: long-haired, dirty, as rank and bedraggled as some injured creature (and injured he was, we soon discovered) found in the woods. Only it wasn't in the woods where my father found him, it was on an icy street in Minneapolis.

Dad was driving in his fuel oil truck, looking for an address to make a delivery, when he saw Gearhardt come out of the side door of a house carrying a coal bucket.

I stuck my head out the truck window and yelled, "Hey Gearhardt!" It was cold as hell that day and all he was wearin' was those ragged coveralls. He looked up, dropped his bucket, and came hopping over and jumped in my truck.

Ma wasn't happy.

"Ruthie, he won't stay. We both know that. He'll be gone when spring comes. Meanwhile, I need a man, and he's a good one—good with cattle. It don't matter that he can't handle the machinery. Ross can do that."

I felt a twinge of pride. Yes, I knew how to operate the machinery now.

"What're we going to pay him?"

"Same as last time—the going rate."

My mother shook her head. "A *hundred* dollars a month?"

"Okay, seventy-five, room and board. He'll work for practically nothing right now. All he wants is a warm place for the winter."

Gearhardt insisted on sleeping in the barn again, where he'd be okay, he said. That first night he carried the blankets and pillow my mother gave him to the cow barn this time, and made a bed in the manger between the calf pen and what had been the horse stalls, where we kept straw to put under the cows and line the gutters. His bed looked kind of cosy. It was certainly warm enough in the barn, if a little drafty, with all the animals in it. And there was the soothing sound of the cattle munching their hay or chewing their cuds, the occasional digestive gurgle from a cow's four-part stomach.

Then Dad noticed Gearhardt cringing a little, favoring one side as he worked, and asked him about it.

"Got hurt," Gearhardt told him.

After some prompting, he told of being knocked down by a hit-and-run driver. Opened his shirt to reveal an infected bruise along his rib cage. Pus was oozing there. He probably had a broken rib or two, besides.

"Good God," my father told him, "that'll have to be looked after! That could kill you, man."

That scared Gearhardt into having his first bath since the last time he'd worked for us and letting my father paint Merthiolate on the wound and bandage it. He took the bath in our washtub behind the closed door of the pantry, after which he cried out,"Ohhh! Ohhh!" like a child, from the sting of the disinfectant.

My mother put his ragged and filthy coveralls in her ringer washer and rummaged through my father's bureau for long underwear and wool outer clothing for him.

After that, there was no excuse for him to sleep in the barn. He moved without protest into the "hired man's room" upstairs. Once again, Dad took him into Hamel for a haircut. And again, as if some kind of foul enchantment had been broken, he was transformed from a dirty tramp into a clean, almost presentable-looking guy who might even attract a woman. That was my father's idea. My mother thought otherwise. "Who would want him with those green teeth? Do you think we could get him to brush?"

We called him a tramp, but he might originally have been a hobo. There was a difference, I would learn from my reading. According to Nels Anderson, author of *The Hobo* (1923) and *Men on the Move* (1940), a hobo, the highest order of homeless men, roamed the country in search of work. A tramp "dreamed and wandered," only working when he had to. Certainly he wasn't a bum. A bum didn't, or couldn't, work because of his health or addictions.

Tramp, hobo, or some of both, Gearhardt probably began his wanderings about the time I was born, during the Depression—he wasn't old enough to have started in the heyday of the bindlestiff, as perhaps our first hired man, Oscar, had. Anyway, he was one of Anderson's men on the move. In the Old West he might have been a mountain man. He might have worked on the lumber rafts that once were floated down the Mississippi; or poled and cordelled keelboats up the Missouri to the fur trade forts. He might have been a prospector. He might have been a cowboy.

January ended, and we were into February, the dead of winter but the shortest month, when you could feel the days getting longer. Then March came with its sunny, windblown days and spring blizzards, winter's last-ditch assaults. The snow was shrinking now, evaporating into the air, seeping into the ground, causing bare and widening circles around the trunks of trees, exposing hillsides, forming puddles. Then it was April and once more, miraculously, all in a few days, the creeks started running and the ice broke up on the lakes. You could feel Gearhardt's restlessness. It was like my own by then.

"Nice day, Mrs. Klatt," he'd say to my mother as she served him lunch at the kitchen table. He spent his days in the barn, but he'd walk up to the house for his noon meal. We kids would be in school, and my father would be in town, working out, so the two of them, my mother and Gearhardt, would be alone on the farm. That didn't scare her.

All those hired men we had. I was alone with them for most of the day when your dad worked out, but I never worried. And I never worried about what they might do to you kids. Nobody worried about such things in those days. I guess we trusted people more. Not like now.

Then one day I came home from school to find Gearhardt gone. Just like the last time.

At lunch he'd said to my mother, "Got to go, Mrs. Klatt."

It was another nice day—there'd been a succession of nice days, so my folks had been expecting this. "When he wants to go, just let'm," my father told my mother. "Pay him up and let him go."

My mother didn't have enough cash in the house, so she wrote him a cheque. He didn't like checques—"People won't cash'm"—but he took hers this time. She told him he could cash it at any of the stores in Hamel.

Then Gearhardt shouldered his pack—his tramp's or hobo's bindle—and started down the road. Though I'd missed seeing him, I knew that long loping stride of his that put the miles behind him. I could see him in my mind's eye going up over the hill just east of our farm and on toward Hamel, toward Minneapolis, toward wherever the road took him. On and on into the outside world I only knew about from books.

Summer

School let out the second week of June, in the green lushness of early summer. I looked forward to summer all winter, and yet the summer "break" from school wasn't the three-month vacation it was—I assumed enviously, contemptuously—for the kids around Lake Minnetonka. *They* spent their summers swimming and boating and sunbathing, I supposed, not making hay and milking cows. *My* break came when school started up again in September.

Called at six each morning by my father to get the cows, I would sit up in bed to see the sun above the horizon and feel the cool freshness of the morning seeping through the screen in my open window. I'd hear the clang of metal from the milk house; with no morning fire to make in the furnace now, my father was already down there, preparing the milking utensils.

I'd dress quickly in rolled-up jeans, a tee-shirt, and my baseball cap, its visor turned up so as not to shade my eyes, and set out along the cowpath for our swamp pasture. I went barefoot to get the cows, still clinging to my Huck Finn conception of myself, stubbornly sticking to boyhood. After morning chores, though, I'd have to put on my work shoes, war-surplus combat boots ordered from Sears or Montgomery ("Monkey") Ward. We sent for a lot of things out of their catalogues. I got my Mossberg twenty-gauge shotgun from either Sears or Monkey Ward when I was fourteen, and later a double-action .22 revolver and a leather holster that I wore sometimes hanging low on my belt and strapped to my right leg, gunfighter style. Both guns were cheap, about twenty-five dollars apiece.

Getting the cows on a summer morning wasn't a bad way to start the day. There was still dew on the grass and birds all around, all singing or calling, all busy with their instinctual lives.

The cows, when I reached them, would start grabbing last mouthfuls of grass. Cows don't have upper incisors, so they can't crop grass like horses or sheep but instead loop their tongues (abrasive as sandpaper) around tufts of it, then deftly snap them off with little jerks of their heads. They hardly chew what they snap up until later when, at rest, they bring up wads (cuds) of partially digested vegetation from the first two of their so-called four stomachs (one stomach, really, composed of four chambers) and only then thoroughly chew it with their molars. I hadn't actually noticed how they grazed until I sat one day and watched them.

Now they were frantically at it, knowing their grazing time was over until after the milking, and I had to yell and wave my arms to get them headed up and moving toward the barn. "Hey! Go home, boss. Yip! Yip!" Eventually I had the lead cow started for home, and the rest, udders swaying, followed her, single file, along the path. I always gave a count then to make sure there wasn't a cow missing. Occasionally a cow might have gotten out of the fenced pasture or, bred by accident too early the previous fall, be off somewhere having her calf. That sometimes happened with first-calf heifers.

Toward fall, you seldom had to get the cows. The grazing wasn't so good then, and they came at your call or lay waiting for you in the barnyard. When I was old enough to do the milking by myself, I learned to call the cows as effectively as my father. You cupped your hands around your mouth and bellowed, "Come boss, come baaasss!" and got an answering bellow from the far reaches of the pasture. Pretty soon, as you were rationing feed before each stanchion in the barn, you'd hear the click of split hooves in the yard and know the cows were home.

When you opened the Dutch door, the cows jostled one another in their eagerness to get into the barn and into their stanchions and at their feed. Some animals went deliberately to another cow's stanchion to steal her feed, but they were butted away by the rightful animal or kicked away by my father or me. Their feed was a base of ground corn and oats with protein and mineral supplements—oil meal, wheat bran, bone meal, flax screenings, various mixtures in varying amounts depending on the season and the cow and whether she was in peak or declining production. Salt, which in the winter we rationed by the handful, was now available to the cows in a block staked out in the barnyard.

In their excitement the cows never failed to empty their bowels, usually splattering the walk on the way to their stanchions and giving me the job of scraping their shit into the gutters with a straightened hoe. In early summer they'd have the "squirts" from the green pasture after a winter of dry feed, and you learned to jump out of the line of fire when a cow in her stanchion raised her tail, heaved, and shot a stream against the barn wall. I'd let it dry there until it could be pried off with the scraper.

We kept a stack of last year's mixed hay beyond the barnyard, where the ground wouldn't be trampled into mud around it, and at night we fed them a few bales of what was left of our fine, third-crop alfalfa. Feeding dry hay to pastured cows gave them roughage and prevented bloat.

We started the milking. Dad carried one of our three Surge machines into the barn from the milk house along with a pail of disinfected water with a rag in it for washing the cows' teats. He turned on the vacuum motor. Its purring caused the cows to let their milk down. Then the stimulation of having their teats washed started them dripping.

I brought in the other two machines, attached the hoses to the pulsators, and carried them with their straps to where my father was squatted next to the first cow in line, easing the teat cups into place. The cups squeezed and pulled at the cow's teats.

Dad stepped out, grabbed another machine, threw the support strap over his shoulder, and stepped between the cows again. The strap had holes at one end and a kind of buckle at the other so that it could be belted around a cow and the milker hung from it. The hose leading from the pulsator was attached to the petcock on the vacuum line and turned on, and the milker, going *shush-shush, shush-shush,* began its rhythmic extraction of milk from the cow.

I went back to the milk house for a pail of cold water for rinsing the teat cups and the two milk pails and returned about the time the first cow was milked out and my father was pulling the machine off her. He emptied the machine into one of the pails and I carried it into the milk house and poured the milk through a strainer into the first of a line of ten-gallon cans. When after several such trips the can was full, I lifted the strainer off, clapped on the can's cover, then heaved the can up and over into the cooling tank. I put fresh pads in the strainer, placed it over the next can to be filled, and threw the used pads, dripping with froth, to our waiting cats. I was doing now what I'd watched my father or Oscar do, years ago, on the Pepin farm. As it was then, milk was all the cats got from us; otherwise they fed on whatever they could catch around the farm, small birds, mice, and rats when they weren't too big and fierce.

A Holstein cow could fill a two-gallon milk pail and then some, though by June, since they were bred so they calved during the winter in the barn, where they could be watched—and helped, if necessary—most of our cows were in declining production. Each cow, as she came in heat after calving, had been bred back, and in the fall as she neared freshening (cows had a nine-month gestation, same as women), she'd be cut off concentrates and wouldn't be milked completely out for a few days, then not milked at all; she'd thus be "dried up" to give her a rest before calving.

We'd be down to about ten cans a day by June, five to a milking. August and September would be our lowest months when we'd ship only six to eight ten-gallon cans a day as compared to fourteen a day during the winter. We reached a peak of eighteen cans a day our last couple of years of farming.

During the milking in summer the barn heated up, grew humid and stifling from the confined animals. I'd get a refreshing breather every time I made a trip to the milk house, but my father stuck it out, maybe allowing himself a break once or twice to go into the milk house and sink his arms into the numbing water in the cooling tank, then splash his face and the back of his neck. When it was really hot, he might strip his shirt off and work half-naked under his bib overalls. *Unless* it was hot, he wore a shirt—and long underwear too, winter and summer, like some character out of a western movie.

Milking the cows in summer when the barn was full of flies, you risked getting whipped in the face by a dirty tail. In a rage, you might grab that tail and try to break it. That could kink a vertebra, put a crook in the tail. You could look down a line of cows—in your or a neighbor's barn—and see a number of such "tell tail" crooks, some animals with a sawtooth series of them, comic (or brutal, if you like) evidence of the poor creature's helpless switching of her tail and the farmer's temporarily insane reaction.

As the cows were milked, I started letting them out of the barn to cool it down. You had to be careful, though, not to let too many cows out at once and so excite those left in their stanchions. They might hold their milk then. Cows liked routine, and could be upset by strange noises in the barn or by the absence of familiar ones, by strange visitors, by loud talking, by sudden moves. When our barn radio, always on during the milking, stopped working, its silence put the cows off production until it was fixed. They missed the radio as much as I did, probably. During the evening milking I sat "glued to the radio," as my father complained, between trips to the milk house, engrossed in *Escape, I Love a Mystery, Suspense, Gunsmoke* (starring William Conrad with his deep, authoritative voice), *Dimension X*.

On summer mornings, as soon as the milking was well started, I'd sneak up to the house, where my mother and younger sisters and brothers were still sleeping, and put water on the stove to boil for drip coffee. Running back and forth repeatedly, I'd eventually have a cup of coffee on the ledge of the milk house window to sip at while dumping milk for my father. It was a somewhat guilty pleasure I kept secret from my father until he caught me at it one day. Then all he said was: "Just be here when I need you."

We pastured our cows on our home eighty as much as possible, only driving them along the lanes to Elder's Forty to prevent overgrazing near home—and only after the morning milking, never at night, to prevent having to go the quarter mile to Elder's Forty for them in the morning. We put the heifers in a separate little pasture between the back of the house and the wooded ridge along our west boundary. When *that* pasture was in danger of overgrazing, we trucked the heifers the three and a half miles to the grandfolks' old pasture along Elm Creek, outside Hamel.

The corn was high enough to cultivate by early June. That first week out of school, after the milking and a quick breakfast—and after I'd cleaned the barn—I might be sent out on our lighter tractor, a used Oliver we had now, in addition to our old Farmall. The Oliver had an electric starter (unlike the Farmall, which you had to crank) and was fitted with hydraulic-lift cultivators that made our old horse-drawn cultivator, with its tongue shortened for hitching to a tractor, obsolete. With the old horse cultivator, you had to stop the tractor at the end or beginning of the rows, hop off, and raise or lower the shovels. With the Oliver you could stay on the tractor and work the levers. Just sitting on it, however, driving back and forth across a field in the heat of summer, hour after hour, was dreamy, dozing work. Yet cultivating corn demanded attention. Unless you kept the tractor strictly between the rows the wheels ran over the corn or the shovels dug it up. Even a moment's lapse could leave a gap in the corn that my father would be sure to spot, and then would bawl me out for it.

Then we were making hay, first-crop alfalfa, cutting and raking and baling it, a section at a time. We would bale a section during the day, and would haul the bales into the barn that evening, never leaving any out overnight, when it might rain. Hauling them out of the fields and stacking them in the barn—going from the heat in the field to the suffocating heat in the mow—was sheer labor, especially for my father and me, neither of us very big. In the field my father hefted the bales up to me on the truck and I lifted and positioned them, making crisscrossed tiers in order to "tie" the load. In the barn he lifted the bales up from the truck to me in the mow, where I stacked them. We did this, load after load through the haying season, until the mow was too high to reach with manpower from the truck. Then the hayfork was used, though the risk was that a bale or two might burst after being dropped from the peak of the barn. When we used the fork, I drove the tractor to work the fork and my father stacked in the mow after each drop.

In some ways, baled hay was harder to put up than loose hay; but it was easier to feed afterwards—easier to pull the bales out to throw down to the cows than to pry loose, snarled, settled hay out of the mow with a pitchfork. And of course you could pack more tonnage into a barn with bales.

But the hay had to be dry, not baled green or wet with morning dew, or the bales would heat up in the barn, bake black inside, and be ruined as fodder. You could almost burn your hands when you stuck them inside a wet or green bale. Hay like that could set a barn on fire.

As for hay that had been rained on, it dried up rather than cured, turned brittle, and the leaves fell off, which is why we never left any in the field. You got only stems from rained-on hay, all right for horses but no good for dairy cattle.

So you gauged the weather and decided when to cut. In the humid heat of those Minnesota summers, with the sun out and temperatures in the high eighties or nineties, you could cut a field in the morning and rake it that afternoon. The next day, assuming the weather had held and the dew hadn't been too heavy, you could start baling by noon.

With luck, you finished by suppertime. Then, after supper and evening chores, you started hauling in the bales. We hauled, as I've said, until they were all in the hot barn—till ten or eleven at night, till after midnight sometimes, under the stars, under flashing meteors, and, in late August occasionally, under a fireworks display of Northern Lights.

In our first year or so of baling, we simply picked up the rows of bales where they'd fallen from the machine. Then my father got the idea of using our stoneboat, pulled behind the baler, to give him a platform on which to stand and stack the bales until he had a pile. Then, driving a five-foot crowbar between the raft-like set of boards, and bracing himself, he caused the bales to slide off onto the field. That left neat stacks rather than scattered, individual bales to pick up afterwards, and reduced the time it took to clear a field of baled hay.

We got help with hauling hay sometimes from visiting town relatives, the man going out to the field with us after supper. Occasionally, when we had hay down in one of our fields outside Hamel and it looked like rain, Dad found a town kid to work with us for a buck an hour. The kid and I would eye each other's shirtless torsos, comparing our muscles.

After first crop haying, usually finished by mid-July, there'd be a lull of a week or so during which we might cultivate the corn once more ("Knee-high by the Fourth of July" was the adage; after that the cultivator started breaking the stalks, and anyway, the corn would be well ahead of the weeds now) and catch up with other work around the place. Then we'd start second-crop, driven by that other old adage, "Make hay while the sun shines." Third crop was the smallest yield, and so late in the summer it conflicted with the oat harvest, but it made the finest hay, hay we fed to the cows in small portions, like dessert, so as to make it last through the winter.

We had two barns, our own on the Mohrmann place and the grandfolks' old barn in Hamel, in which we stored hay grown on the land we rented and later bought from them. In its two fields, separated by the pasture along Elm Creek, we raised hay or corn or oats, in rotation. Nobody could have guessed then, least of all my father, that those forty acres, together with the adjoining twenty bought from my uncle Lucien, would eventually become the golf course my father built after he quit farming.

The weather. The weather in summer during the growing season. A farmer depends on it, endures it, curses it when, as often happens, it turns against him. A cloudburst can flood a field, flatten or "lodge" great swatches of grain, make gullies in one place and leave deltas of eroded soil in another. Wind, not to mention tornadoes, can lift shingles off buildings or raise entire roofs. Lightning can kill your livestock. Hail can strip, shred, pound your crops into the ground, as well as break windows and dent your car. Drought. It cracks the soil, withers the crops, dries the grass from green to yellow in the pastures. Some of this happened on our farm, but nothing so devastating as to ruin us. We were too diverse for that. Dairying was our mainstay, but we weren't dependent, like our relatives in North Dakota, on wheat, their one crop.

One morning I got up for chores and looked out my bedroom window to see a horse in our front pasture. It was a little buckskin, like an Indian pony, cropping grass out there by itself because our cows were in another pasture. It raised its blunt head to look at me as I came out of the house, then returned to its grazing. Seeing a horse in our pasture was a novelty now. We'd sold our work horses not long after the war.

"There's a horse out in our pasture!" I told my father excitedly. He was putting out feed in the manger.

"Never mind that now. Just get the cows."

After I'd brought the cows home and they'd filed into the barn and were stanchioned, I said, "Look, Daddy. There's that horse I was telling you about."

He looked out the open top half of the Dutch door. "Well I'll be. Must be one of Reichert's."

Reichert had a rendering plant on the old Huar place, a mile or so away, where he butchered horses for dog food, we'd heard, old draft horses like the ones we'd gotten rid of, and wild mustangs, scrubby, ornery little hammerheads caught out in the western badlands and trucked to his plant. This looked to be one of those.

After the milking, Dad called Reichert, who said over the phone that, yup, the little buckskin sounded like one of his, run away before he could shoot it.

We could have it, he said, for twenty dollars. That would save him the trouble of coming over to get it.

"You want a horse?" Dad asked me.

We got him into the barn somehow and tied him up in the old horse stall. And, yes, he was a stubby, mean little mustang, a stallion, a kicker and biter. I fed him and tried petting him. He tossed his head up, laid his ears back and showed his teeth. He was more than I could hope to handle until he was broken.

We knew of somebody who might do the job, a Hamel kid whose banker father kept horses and who'd been riding since he could walk. Jody was only a couple of years older than me, but he'd ridden broncs in rodeos and knew how to train horses and even break them.

We called him up. A day or so later he rode out to our place. He looked my horse over, got a training halter on him, and tried leading him around. It wasn't easy. "It'll take a couple of weeks," Jody told us. "What're you calling him?"

"Buck," I said. "He's a buckskin, isn't he?"

"Buck's a good name," Jody said. He mounted his horse and led Buck away to his folks' "ranch" outside Hamel.

While Jody had Buck, Dad and I looked through the Sears, Roebuck catalogue and found a plain, workmanlike stock saddle to send for. We sent for a bridle, too, and a saddle blanket, and cowboy boots.

Some ten days later I saw a horseman, leading a second horse, approaching the farm. It was Jody, with Buck.

"Here's your horse," he said. "He's green broke, so you'll have to ride him every day."

Buck still had a wild look in his eyes. When I moved to pet him, his ears went back, but he didn't swing around to kick at me, and he didn't bite.

"He's a feisty little bronc," Jody said. "You got a saddle?"

He helped me saddle Buck, showing me how to wait until the horse exhaled before tightening the girth, then helped me ease myself onto him. "Take him real slow," Jody said, "until he's used to you. He's a good little horse. You'll have a lot of fun with him."

Up to now I'd ridden only work horses, which had always seemed indifferent to my slight weight on their broad backs. This animal pranced and sidestepped under me, as if he wanted me off. But my legs wrapped nicely around him. He was just my size. I felt like a cowboy.

"He neck-reins," Jody told me. "And I've taught him to stand when the reins are on the ground."

He climbed onto his own horse. "No, don't ride along with me," he said. "He's used to my horse and you might not be able to turn him. Hold him till I'm out of sight, then ride in the opposite direction. Give him a run."

Buck liked to run. I hung onto the saddle horn and tightened my knees against him, bouncing hard, almost pitching off. Eventually I learned to relax in the stirrups, rock with the horse's movements. Meanwhile I was stiff and sore those first few days, then gradually hardened up. Pretty soon I had a rider's calluses on my ass.

I forget how much the breaking cost. Jody charged, what, twenty-five dollars a week? So to have Buck "green-broke" may have cost fifty dollars, twice my monthly wage then. Add to that Buck's initial cost and the cost of his saddle and bridle and the boots I wore. Did my dad take any of that out of my wages? I doubt it. More probably, as was his way, he got a promise from me to work harder.

Through the rest of that summer of 1950 and into fall, I rode Buck faithfully, every day, and gloried in having a horse. Jody had taught him to come to you in the pasture for a sugar cube or an apple. That allowed you to slip a halter over his head and lead him to the barn to be saddled. I never tried to ride him bareback. He was too balky for that. I didn't want to get thrown, but I was thrown, and before that I fell off him once because I'd failed to saddle him properly.

It was a day soon after I got him and I was using him to round up heifers. I thought I'd tightened the saddle girth enough, but as he dodged this way and that at just a touch of the reins (he was an instinctive cow pony, able to herd cattle as a dog does sheep), the saddle slipped over his flank and under his belly and I rolled onto the ground. Buck ran off, then stopped when the reins fell in front of him. He let me approach and mount him again after I'd cinched the saddle up right this time.

Then one Saturday in the middle of winter when, because of the cold, I'd neglected to ride him during the week, Buck bucked me off. He was ornery as I saddled him, and I had a hard time getting on him. As we started out of the yard, I thought, *Okay, you little bastard, let's give you a run.* I eased up on his reins and said "Haw!"—his signal to gallop. Instead he put his head down and started bucking. *Whump, whump, whump,* and I sailed over the saddle, turned a somersault and landed, still holding the reins, in the snow. I was jerked upright as Buck reared away and I held on. He stopped dead then, trembling, his ears back. *If you ever get thrown, get right back on him,* Jody had said. I was scared and Buck kept turning as I tried to mount him, but I got on, finally, and pulled his head up. We had a good, nervous ride after that.

I took to wearing my cowboy boots to high school, clumping down the halls, imagining my legs were bowed like some wrangler's out of a Zane Grey western. The boots gave me a certain style, I thought, and one or two girls gave me a look.

We got our first television set the next summer, a sixteen-inch Motorola. After that I could watch old westerns and such early TV series as *Range Rider* and *The Cisco Kid,* then act out the hard riding they featured on my own horse.

Nicky Herman had a horse, too, that year, and we sometimes rode together. Once we tried riding at each other, like jousting knights, and nearly tore our legs off when we collided. That made me wonder how often that had happened in the days when men charged at each other on horseback.

Gradually the novelty of having a horse wore off. I rode him less and less. There was too much work to do and too little time for pleasure riding; our place was a dairy farm, not a cattle ranch, as my father kept reminding me, so having a riding horse wasn't practical.

Because I wasn't riding him enough, Buck got mean and unruly again. I learned to hold the reins tight under his chin when leading him; otherwise he'd move up and bite my shoulder. Saddling him became a challenge; he'd try to lie down and roll on the saddle. Then, mounted, you fought to control him while he went through his bag of tricks: mock shying, little jumps and half-bucks. Galloping down the road, he'd suddenly veer toward a telephone pole to try to scrape you off. Finally, one day, I couldn't get on him. Dad walked over and said, "Here, lemme show you how to ride that fucker."

He swung into the saddle and rode Buck into a plowed field. Around and around he rode, plunging the horse through the soft, broken ground until he was heaving and lathered.

"There," Dad said, dismounting and handing me the reins. "Try riding him now."

Pastured with the heifers, Buck started chasing them, biting and kicking. One day he ran a heifer up against the fence and broke both her front legs.

"That's it," Dad said. "We're getting rid of that mean bastard."

We sold him to a farm kid outside Hamel for about a hundred dollars, saddle and all, losing money on the deal. "Good riddance," Dad said.

Of course I missed my little mustang as soon as he was gone, but I was relieved, too, I had to admit to myself. He'd become just another chore, finally, more work than fun. I was happier, after all, with my fantasies.

I was lying naked on my bed after a swim in Lake Independence one night, playing with myself. (This was in the summer after my first year of high school, before I got Buck.) I felt good after the swim, clean and cool, and the sheets felt good on my skin and my stiff penis felt good in my hand. I'd been fondling myself since childhood, taking mild, diffuse pleasure in it. Lately the sensation had grown sharper, more deeply pleasurable.

It was very pleasurable that night and I kept at it until the sensation began to gather and then to pull and squeeze at me as my hand seemed to work of itself—until abruptly there was an intense, aching, flooding release of something sticky on my belly.

I pulled the light on over the bed. Yes. There it was, that milky, gooey stuff of wet dreams. Only I'd achieved it while wide awake. I'd jacked off!

That first time was an accident. The second time, only minutes later, wasn't.

The confessional window slid open and Father Michael's silhouetted head appeared on his side of the screen.

"Bless me, Father, for I have sinned. My last confession was a couple of weeks ago (a bald lie; it had been at least a month, but I tended now to cut the time in half). I talked back to my parents. I was mean to my sisters and brothers. I had bad thoughts."

I took a deep breath while the priest waited, his ear to the screen.

"And I sinned against purity with myself, Father."

He didn't ask what I meant; seemed to know exactly what it was, only said, "How many times, my son?"

I hadn't expected that. I thought back wildly: once, more often twice, every day for how many days since that first night after swimming? The figure was monstrous. I halved, quartered it.

"Six times, Father."

He knew who I was. Father Michael knew all his parishioners, in and out of the confessional. And you always knew from his silence behind the screen while you recited your sins, from his consoling or scolding words afterwards, from his patient or exasperated tone of voice, exactly how he felt about your human weakness.

"That all, my son?"

"Yes, Father."

I heard the squeak of his wicker chair as he repositioned himself behind the screen. I heard his sigh.

"Listen, you're a young man now. With a young man's urges. Someday you'll find a woman you love and who loves you, and you'll be married. Then the urges you feel now will have their lawful outlet. Those urges were put there by God for procreation. You know what that means?"

"Yes, Father."

"Having children. Having children in the state of Holy Matrimony. Until you've reached that state, until you're married to the woman who will be your wife for so long as you both shall live, *you're just going to have to keep it in your pants.* For your penance say ten Our Fathers and ten Hail Marys. Now make a good Act of Contrition."

I said a fervent Act of Contrition while Father Michael intoned in Latin, after which I was told to go and sin no more. Father slid his window shut and I stepped out of the confessional and into the quiet church. There were a dozen or so people in the pews, their heads bowed, mulling over their sins. Three or four others were lined up a discreet distance from the cabinet-like confessional, waiting their turn. The next penitent stepped forward. It was Willie Tyson, a junior in high school now.

He smirked as we passed each other in the aisle.

Winter

In the cold dark of a winter morning, I hear my father get up. I hear the creak of the bed downstairs as he leaves my sleeping mother, the soft click as he closes their bedroom door and moves into the dining room. The thump of his bad foot as he walks below me, almost directly under the floor register in my room.

I hear the whisper of cloth as he dresses himself. Grunts as he sits in the chair outside the bedroom and struggles into his work shoes, the one shaped normally and the other, I know, bent and cracked at the instep because of the loss of half that foot to the silo filler. I hear a groan or two—his bad foot must be aching—and a resounding morning fart as, upstairs, my eyes still tightly closed, I try for a few more seconds of sleep.

Then the stairway door opens and quietly, so as not to wake my sisters in their rooms next to mine, my father calls up to me.

"Ross. *Oh* Ross."

"Yuh!"

I open my eyes to the cold dark, reach for the string hanging over my bed and pull on the light to see my breath under the naked bulb in the ceiling. When I roll out of bed the cold wooden floor feels hot against my bare feet. I grab my long underwear, wool socks, jeans, flannel shirt (I sleep naked, winter and summer, despite my mother's anxiety about it: *What if you died in the night?*), and clump down the stairs.

It's a little past six by the kitchen clock above the refrigerator. I dress in front of the heat register in the kitchen, though this early in the morning there's never any heat issuing from it. The folks, afraid of coal gas, let the fire go out at night (they'd read of a whole farm family asphyxiated once, and indeed there was a night when our furnace backed up and we kids were jolted awake in the smoky house and pushed outside into the sub-zero cold), so the only thing coming up the register as I stand there shivering are the sounds of my father shaking down the clinkers in the furnace and whiffs of wood and coal smoke as he gets the fire going. It'll be much warmer in the barn this time of morning—one reason to hurry and get down there—and a good hour before the house warms up. That'll be about the time I come up from the barn to get ready for school.

Dad comes up the cellar stairs, says, "C'mon, let's get goan," and we step out on the porch, where our barn clothes air on hooks. We pull on our overalls and then our jackets, our wool caps, our buckle overshoes. Stuff our pants into them, then buckle them up.

Then we step off the porch onto the packed-snow path to the barn. The stars shine coldly overhead. You can tell when it's below zero Fahrenheit by the sting in your nose, the grip of the air on your face, and the squeaking of the snow under your overshoes. Above zero, the snow crunches underfoot.

Dad opens the barn door and steam gushes out; inside, the humid animal warmth strikes you like a damp towel pressed to your face.

I open the door to the barnyard and more steam escapes into the cold. I pull the manure carrier in on its overhead track, then latch both halves of the door again. There are two long gutters and one short one to shovel out, behind the twenty-eight cows we're milking now, while Dad feeds the cows their rations of ground feed and starts the milking. I have the barn to clean (three carrier loads winched up and pushed outside to add to the growing pile of manure in the yard), then fresh straw to spread in the gutters and finally hay to throw down from the mow and spread in the manger for the cows to munch on. I have to do this, then get up to the house in time to wash, change clothes, and eat before the school bus stops at the head of our short driveway, about 7:30.

The barn drips with moisture. All those breathing animals, maybe forty of them counting the weaned calves in their pen and the suckling calves tied behind the cows, all confined in the low walk-in basement of the barn and contributing to the condensation. There's condensation from the stinking muck in the gutters and a reek of ammonia in the corners so powerful it makes your eyes water, while up in the cold, dry mow there's the spicy smell of cured alfalfa and a marvel of hoar frost encrusting the rafters above the haychute. It's evidence of the animal warmth below, a white fantasy, like an illustration in a book of Nordic folk tales.

❧

Dad still worked out in the winters, delivering fuel oil in Minneapolis, so it was up to me to start the chores when I got home from school. One winter he worked as a salesman for an old friend's Ford dealership in Wayzata, but that bored him, seemed a lazy man's waste of time. *I had to get into a suit every day, like I was going to church, and all we did was sit around the showroom talking, waiting for prospects.* Then, for an elected two-year term starting in the fall of 1950, he served as tax assessor for Medina Township.

It was after five when Joyce and I stepped off the bus from Mound. Joyce would change clothes and help Ma start supper while I skimmed the front page of the Minneapolis *Tribune* for news of the "police action" in Korea.

"Better get going," Ma said. "Your dad'll be home pretty soon and he'll be mad if you haven't started chores."

"Yeah, yeah."

I'd tank up on milk and whatever sweets Ma had baked (if I hadn't filled up already on the goods I could buy in the bakery across from the school and eat on the bus). Finally I'd change into my work clothes and walk down to the barn.

The barn had to be cleaned again. Usually I was still at it when Dad's rap on the door told me he was home. I'd unlatch it and let him in. He'd go right to feeding the cattle their ground feed. I'd finish shoveling out the gutters, line them with fresh straw, then feed silage to the cows while Dad disinfected the utensils and set them up in the milk house. Then we'd walk up to the house for supper.

If Dad was late getting home, as he sometimes was, I did the pre-milking chores on my own, and would be in the house already, eating my supper, when his car lights turned into the yard. Then he'd give me a nod of approval as he stepped into the house. That was my reward.

After supper we did the milking together. In the dead of winter with all the cows producing, Dad operated our three Surge milkers while I threw bales of hay down from the mow onto the manger, then broke and spread them for the cows to munch on. I had calves to feed, livestock to bed down and, between these jobs, the pails of milk to dump into the ten-gallon cans in the milkhouse that my father kept filling. It would be ten or ten-thirty at night before we finished.

Up at the house it was bedtime by then, but stubbornly, we'd stay up a while, my father to read the paper or go over accounts with my mother, myself to read a little or listen to the rest of a radio program I'd heard the start of in the barn. I'd fall asleep sometimes, sitting on the couch in the dining room. My younger sisters and brothers would have been in bed by nine o'clock.

During the milking, Dad noticed which cows might be in heat. You could tell by the swollen look of a cow's vulva or the nervous switching of her tail, but the only sure way of knowing was to let the bull loose to range sniffing past the lines of stanchioned cows until he stopped behind one in heat, flared his nose so that the ring in it flipped upwards, then mounted her. Occasionally that's all it took: the bull's rodlike penis unsheathed and he rammed it home. Usually, though, he needed help. The Holstein is a big cow, and a full-grown Holstein bull is a two thousand pound beast that can find it difficult to raise his bulk to the height of a cow in stanchion.

So here's what we did. Dad went around to the manger and forced the cow to step back into the gutter. That lowered her behind. I, meanwhile, squeezed next to the cow and pulled her tail back, offering a clear target for the bull. The bull mounted and I arched away like a matador to avoid his thrust.

Not very scientific, but it worked. Except once, the bull's ardor was such that he completely missed the cow and I had the singular (and, later, humorously recounted) experience of being "bred" in the armpit.

A snowy night. Dad was late and I was still in the barn when I heard his car drive into the yard. I waited for his rap on the barn door, but it didn't come. I fed the cows grain and washed the milking utensils. Then I walked up to the house for supper to find people standing around in the kitchen: Dad, with another man whom I recognized as the guy from Missouri who owned a service garage outside Hamel; Ma and my sisters, in tears; everybody staring at my father.

His back was to me as I walked in. Then he turned around and I saw his face. "Your dad's been beaten up!" Ma said fiercely, which was hardly necessary because I could see the evidence for myself. His face was discolored, swollen almost out of recognition, his eyes puffy, slitted, oozing blood. When he talked, it was as if he had mush in his mouth.

In angry bursts, I was told what had happened.

Driving home from Minneapolis, my father had cut over from Highway 55 to the Old Rockford Road. As he turned onto the road, lights flashed in his rearview mirror and he looked up to see another car come over the hill behind him, spin out of control on the snowy road, and wind up "backasswards" in the ditch.

"So I stopped to help the guy out," my father said. "I shoulda kept going."

The driver of the other vehicle was a young war veteran from Hamel who, as Dad approached him, started spewing profanity.

"The *fuck* you learn to drive? Don't you stop before you turn onto a road?"

Perhaps my father hadn't stopped, just looked to his left before turning right onto the Rockford Road. But the man's yapping, inciting belligerence instantly triggered my father's.

"Hey," he told the guy. "*Hey.* You were just driving too goddam fast."

"Yeah? Yeah? Well, *you* were driving like some fucking old woman!"

That caused my father to say, "Look. You wanna fight, asshole, or you want some help outa the ditch?"

The guy wore glasses. Now he pulled them off, put up his fists.

"All right," Dad said. "If that's the way you want it." It was a stupid situation, he knew, but what man could back down from it?

He started pulling off his jacket, momentarily entangling his arms, which is when the son of a bitch stepped in and started punching. My father went down. Then the asshole jumped on him, slammed his head against the pavement—*Good thing there was snow on the road!*—and tried to gouge his eyes out. Where'd he learn that? In the Army maybe. Across the road, from inside his house, the Missouri man heard yelling and stepped out to see the cursing, struggling figures on the road. He ran over and pulled the man—*He was acting crazy!*—off my father. Then, since Dad couldn't see to drive, he got into Dad's car with him and drove him home.

"What about the other guy?"

"We left him there," Dad told me. "I hope the bastard's still stuck."

Did Ma drive the Missouri man home, that Good Samaritan? Probably. Anyway, despite the fact that he could hardly see, Dad went down to the barn with me and we did the milking.

The following Sunday, after church and dinner at the grandfolks', my father and my uncle Gene went to see the guy—specifically to see the young guy's parents. He was unmarried, still lived with his folks. Showing his bruised face to the family at their own Sunday dinner, what did my father expect? Sympathy? Some kind of compensation? Maybe he just wanted the guy's folks to see what their boy was capable of. They laughed at him.

"Oh, they're a tribe," Uncle Gene told us. (Not that we weren't a tribe ourselves.)

"Hell with'm," Dad said.

I wanted to know what the guy looked like. I wanted to know so I could beat him up like he'd beaten my father. But when he was pointed out to me one day, a stocky, muscular young man maybe ten years older than I was, whose glasses didn't make him look like a sissy, I knew I'd have to wait a while.

I never faced him, of course. I never even spoke to him. Eventually he moved away.

I was a sophomore in high school that year and signed up, as I had my freshman year, for wrestling. The first day of practice, I went down to the locker room and found posted on the coach's blackboard the date and place of the team's first meet, and the names, opposite the various weight classes, of those who would wrestle. My name was on the board as the team's ninety-five pounder.

I still weighed less than ninety-five pounds. Consequently, I was told to eat before the weigh-in, rather than starve. "Hamburgers and malts," the coach said. "Fatten up!"

The meet was with Robbinsdale, a big school in northwest suburban Minneapolis with a tough wrestling team—one of Mound's great rivals.

The night of the meet, I didn't go home after school (Dad had excused me, somewhat grumpily, from my farm chores) but stayed with the team for the weigh-in, ate lightly with them at a diner in Mound afterwards, then rode in a school bus with Coach and the team to Robbinsdale.

All the way there, I was sick with nervous anticipation. How would I do against my opponent, whoever he turned out to be? I was scared, in fact—scared sick of losing.

In the Robbinsdale gym, the boy I would wrestle and I stripped off our team jerseys and crouched, sizing each other up from opposite sides of the mat. He was slightly bigger than me, and for a town kid he looked strong. For a farm kid, I must have looked pretty skinny.

At the call, "Wrestle!" I danced out, ducked and grabbed his legs, lifted and took him down. Two points. He twisted frantically, got to his knees, whirled to face me, but I followed him around and got behind him. Two more points. A cheer went up from my teammates.

Then my lightness and inexperience betrayed me. My opponent raised his backside, lifting me off the mat as I stupidly hung on. (I might have let go, allowing him to escape to face me, given him the point—or dropped back to pin his legs and stay behind him.) He threw me off, got on top of me, and at once achieved a pin hold. I struggled under his weight, threw my head back, dug in my heels and pushed, trying to lift myself and twist over onto my belly. The referee slapped the mat. Pinned! After hardly a minute into the first period.

I stood up, humiliated, my neck feeling broken, and looked over at Coach. He looked away. My teammates looked away.

At the next practice, I had an elimination match—and was eliminated. During the intensity of the match, I experienced, for maybe the first time, that sudden cold plunge into total consciousness, that shock into the here and now, as if snapping awake from a dream. I struggled with all my might but knew I was going to lose, knew I would be defeated. And of course I was.

Coach seemed sympathetic. "Take a long shower, hot as you can stand," he told me. "Soak those muscles. Then work your towel across your neck and shoulders, *hard*." There was the faintest scorn in his voice when he emphasized the word "hard."

Standing alone in the shower, my ex-teammates still working out upstairs, I let self-pity wash over me with the near-scalding water. It was soothing, a relief to know that I was off the team now and could give up trying and settle back to just practice, to watching others strive with all their might on the wrestling mat. "You need heart," Coach reminded us. Instead I had a sense of failure already familiar by then and darkly pleasurable, born partly of adolescent angst, I suppose, but mostly of my growing sense of entrapment on the farm and my hopeless-seeming struggle with my father, who raged at me out of his anger and frustration, raged at my dreamy lack of interest in farming.

"You're worthless!" he'd yell at me. "You're never going to amount to anything!"

Eventually I could hear someone telling *him* that when he was a kid.

Late that winter there was a wrestling tournament in gym class. I signed up for it, pretty sure I could beat out any of the amateurs or second-stringers I'd face.

There were only two other guys in my weight class. One was a small, puny looking kid who never went out for sports (why was he wrestling now? For reasons like my own, maybe), and who, according to locker room gossip, "did it" with his sister. I thought of that as we faced each other. At the command to wrestle, I charged across the mat, grappled with him, and took him down. He was weak as a girl, I found, probably no stronger than his sister. Revulsion for him filled me with power and I pinned him, as I'd been pinned at Robbinsdale, within seconds of the first period. He struggled up afterwards and gave me a limp, soft handshake. I felt like a bully.

My second match was with a fellow B-team wrestler who, though he wasn't a farm kid, I knew to be tough. We wrestled two periods during which I avoided being pinned and made points. Then, in the third and last period, after losing the coin toss and letting him take the down position, I blocked his tired attempt at a sitdown and breakaway, stayed behind him, held him, and rode him out to the end of the period. I won on points.

It was two-bit glory, but I basked in it.

March came in like a lamb, as they say, with balmy winds and thawing snow. The snow contracted, the dirt roads turned muddy, and the low places filled with meltwater. It was another spring, and I came down with another bout of spring fever.

> *My heart knows*
> *what the wild goose knows*
> *And my heart goes*
> *where the wild goose goes*

That hit song of the period, in Frankie Laine's stirring voice, expressed my own wild yearning.

I reached my maximum height about this time, five-foot-five in my work shoes, though I carried the hope into my twenties of growing another inch or two. Some girls thought me "cute," maybe, which wasn't the same as "cool" or "sexy." Occasionally I'd get a hint from a girl that maybe she liked me "that way", as my sister Joyce called it. But the girls I liked "that way," girls who seemed as unobtainable to me as movie stars—girls you'd like to "breed," as another farm boy put it—the "queens" or those known to be wild or tough, they always seemed to go for the jocks or the hoods, guys who either were shining athletes or greasy-haired bad. Me? I was just a farm kid, stinking of the barn, with a boy's body and a baby face.

I had yet to ask a girl out for a date, though at a New Year's Eve party that winter I'd received my first kiss from a girl. The utter thrill of it! And the next morning after the milking, sleepy and hung over, standing on our bobsled and forking cowshit over a snowy field in below-zero weather, my lips still burned with the sensation of her kiss (who was she? Some girl I'd hardly noticed before). More than the work of spreading manure by hand, the lingering feel of that girl's sweet lips on mine kept me warm that cold morning.

In May there was a track meet at school that coincided with regular practice. Instead of practice, the coach had nothing in mind for the B squad except to warm up with the A squad and then attend the meet.

The dedicated and conscientious put on their jock straps and sweat clothes for the warmup run around the track before sitting on the bleachers to cheer our athletes. Others of us, myself, Hal, Dan, and a farm kid from around Loretto, didn't bother. Instead, after lagging behind in the locker room, we slipped out and started walking toward home.

A couple of miles out of Mound, thinking it was safe now, we started hitchhiking. We were picked up by a man none of us recognized but who turned out to be a junior high instructor, new on the staff, who allowed that our skipping school was no concern of his since he was doing the same thing himself. He drove us to Lake Independence, where we asked to be let off.

We got off there because I'd talked the others into renting a boat and rowing out to the island in the lake. It had stood out there, off the far shore, calling to me, since the *idea* of islands had taken hold of me after reading *Tom Sawyer* and *Huckleberry Finn*.

The island was unnamed, so far as we knew, and uninhabited, though in summer it was part of the youth camp across the lake. This early in the season, we figured, it would be deserted.

At Independence Beach we rented a rowboat and, taking turns at the oars, headed for the island about a mile away across the lake. While the others in the boat chattered, I was speechless with excitement, imagining us heading for Jackson's Island in the Mississippi or some island in the South Seas.

Approaching it, we saw that the island was really two islands. Stretching along the outer shore of the main island, like a protective reef, was a low, narrow bar covered with willows. We rowed around it to the main island and found a marshy cove in which there were mounds of mud and cattails. They were muskrat houses. Instantly I was my miniature, imaginary self, Kianga, and saw a village of miniature lake dwellers or South Sea Islanders hidden away on this island that was a lost little world, one Kianga might discover and where he might meet a beautiful native girl (Debra Paget in *Bird of Paradise,* a movie I'd see that summer, became her embodiment) and have sex with her and live with her happily ever after.

We pulled into the cove's mucky shore, tied the boat to a tree, and climbed an overgrown path up a steep bank to the top of the island. The trees changed from willow and cottonwood near the water to basswood, maple and oak, higher up. Grapevines draped the trees along the water. On the island's plateau, a well-worn path took us around it. There was a dock at the other end of the island, across from the youth camp, and on its lee side, where the island was separated from the marshy lakeshore by a channel only a hundred feet or so across, there were the remains of an old bridge. Rotting wooden posts stuck up in the channel and continued, in a broken line, across the marsh to the higher ground beyond.

We headed into the center of the island, swatting mosquitoes and deer flies, pushing through tangled, second-growth woods, and broke out into an overgrown clearing, crawling with garter snakes, in which we found the ruins of a small house. There was just the basement left, a rectangular, rock-lined hole strewn with splintered and weathered boards. Beyond it was another hole, a collapsed well, and at the far end of the clearing were a few scraggly fruit trees.

Leaving Kianga for the real world now, my mind began to swim. To live here, in my own little house, with an orchard and garden. To be self-sustained and self-contained here. To have this small island, this detached, circumscribed little world, surrounded by nature!

Rowing away from the island, we found ourselves infested with wood ticks. They were everywhere, under our jeans, up our legs, in our pubic hair, flat, faintly tickling little vermin, some already fastened to the skin and sucking blood. "Jesus!" Hal said. "I hate ticks," and started pulling off his jeans, rocking the boat.

I kept my mouth shut. I didn't like ticks either, but they were part of the island, part of the natural world into which, now that it was spring again, I wanted to lose myself.

4

This 120-Acre Farm

Escapes

At the supper table one evening my father started bawling me out for my daydreaming ways.

"What in the hell do I have to *do*, for chrissake, for you to take an interest?"

It was early June of 1951. School had let out the week before. It was the start of summer, the start of first-crop alfalfa. I was thinking about the summer and how fast it would go and how much of it would be lost to farm work.

"*Listen*, goddammit!"

I looked down as he ranted on about my not giving a damn and how I wasn't a kid no more and it was time I got the lead out and pulled my head out of my ass and *paid attention!*

I stared at the food on my plate, unappetizing now and getting cold. Nobody else said anything. Little Nancy sat quietly in her high chair, though a moment before she'd been entertaining us with babytalk. The twins, when I glanced their way, looked glad it wasn't *them* under their daddy's fire. My sisters were impassive. Ma had that helpless look she got whenever my father started swearing. She knew better than to speak up for me. Later, maybe, but not now.

Then I realized my father had stopped talking. I looked up. There was a funny little smile on his face.

"You work good this summer"—he was mumbling now—"and I'll let you take the car up north fishin'. Give ya most of a week. Give ya twenty-five bucks spending money. You can take your buddies along. Whattaya think of that?"

I worked good. We finished first-crop alfalfa by the end of the first week in July, and I spent the weekend preparing for my trip. Early the next Monday, sitting behind the wheel of the folks' new '51 Ford that I'd filled with gas from the hundred-gallon storage tank we kept mainly for our tractors, the trunk packed with canned goods and my fishing and camping gear, and with the promised twenty-five dollars in my billfold and my sisters' envious looks following me down the driveway, I drove to Hal's place to pick him up, and then to Loretto to pick up Dan. "Going up north, *by themselves,* and the oldest only sixteen!" Dan's mother had remarked to my mother one Sunday after church. "I don't know . . ." But Dan's folks, and Hal's too, let their boys go with me.

We headed up Highway 55 on that glorious July day, all shouting, all talking at once. We'd gone camping before together, twice, in the woods of Elder's Forty the last couple of summers, but only for a night each time. This would be for almost a week. The Three Musketeers! All for one, and one for all!

Past Maple Lake, driving parallel with the Soo Line tracks, we raced a freight train trundling along at about fifty miles per hour. We drew abreast of a boxcar with two young hoboes sitting on top of it. I honked, and one of them waved. He sat hunched on the walkway along the top of the car, his back to the wind, black hair blowing in it. A few years later, after reading *On the Road*, I would wonder: was that Kerouac? Whoever it was that day, his riding up there on the windy, swaying top of that train looked more uncomfortable than romantic.

Our hazy destination was Alexandria, Minnesota, a resort and farming community about a hundred miles northwest of the Twin Cities. But when we got there, a couple of hours later, it looked too settled and pastoral—too much like home, in fact—and anyhow, it wasn't far enough north. So we drove on, another hundred miles, to Detroit Lakes.

In the photo album I've kept of that and later trips, there are snapshots of us (two in a picture, the unseen third aiming my Brownie Hawkeye) at various points of interest along the way. There's a picture of Hal and me, leaning against the sign marking the town of Belgrade, under which I've written, "Happy and carefree." There's another of just the highway stretching away to the horizon in a flat, virtually empty landscape where distant clumps of trees signal the presence of scattered farms. We were out on what had been unbroken tall-grass prairie in the old days, once buffalo country, on a northwest course roughly following the old Red River Trail along which, before the railroads supplanted it, slow ox carts squeaked back and forth between frontier St. Paul and the Red River Colony in Manitoba.

At Fergus Falls, we swung due north to Detroit Lakes.

We got there after five hours on the road and drove through town, then out into the country, looking for a campsite. Every resort on the main highway was filled, it seemed—this was July, after all, and people were up from the Cities, from Chicago even. Finally, on a secondary road, well off the main highway, we came on a sign, MAPLEWOOD VIEW RESORT, LONG LAKE, that promised "boats, baits, cabins, free picnic grounds."

We drove down through trees on a dirt track with grass growing between the wheel ruts to the lake. A beefy, middle-aged man with a pleasant face walked up from the wooden dock. He was in rolled-down hip boots, had the sleeves of his shirt rolled up to his elbows, and wore an old fedora on his head.

"Hel-lo," he said with a soft Scandinavian lilt. "What can I do for you boys?"

We asked if we could camp.

"Yah. Where you boys from?"

His name was Gilbert Benson. "Yust call me Gilly." He was a smiling old bachelor, an ex-farm boy. Since he wasn't that busy—we were the only customers in sight—sure, we could camp on his place, as long as we liked. How much could we afford to pay?

We settled, I think, on a dollar a day. For fishing, we could rent one of his old flat-bottom rowboats for two bits an hour. There were three or four of them overturned on the shore, and one new-looking, round-bottom boat in the water, tied to the wooden dock. It had a metal stern plate to which you could attach an outboard motor and, yes, Gilly had an outboard to rent. That boat rented for seventy-five cents an hour, and the motor cost a dollar an hour: too much, we decided.

Up the bank from the dock, tipped over and wedged between two trees, was Gilly's fish house, hauled to shore before last winter's ice left the lake. In the clearing overlooking the lake were a couple of picnic tables, and off in the woods was an outhouse. Beyond that were a couple of cabins, empty the whole week we were there.

Postcard, July 9, 1951:

> *Dear Ma & Dad*
> *Well we got here in one peace [sic].*
> *We had a heck of a time finding a place*
> *to camp. We're in town now buying food.*
> *I'll see you Thursday.*

We were a hundred and fifty miles south of Canada, not way up north, but there were lots of lakes and mixed birch and pine woods all around, and the resiny *smell* of up north.

That evening a pair of locals showed up, a thin man in farm clothes lugging his own outboard, and his overweight wife in a print dress carrying their rods and tackle box. The man attached the motor to the round-bottom, yanked it to life, and the couple buzzed off.

The rest of the week, except for the old Swede's occasional presence and an evening fisherman or two, we had the place to ourselves.

The old snapshots show us unloading the car, setting up my pup tent, skylarking around camp. I'm shirtless, tanned dark, in rolled-up jeans and barefoot, Huck Finn and Tom Sawyer combined. There are pictures of us from later in the week holding strings of sunfish and crappies, caught in the lake. We fished every day, casting from the dock or rowing out in one of Gilly's boats in the late afternoons or early evenings.

It was on this trip, I think, in a drugstore in Detroit Lakes, that I discovered *Blue Book*, one of the last and greatest of the old pulp magazines. Through lazy mornings in camp, I read the issue I'd bought from cover to cover. It was full of men's action and adventure stories that I gobbled up as if they were food. They were the kind of stories I wished I could write.

Then the four days my father had given me were over. We packed up, said goodbye to Gilly, and headed home, not talking much, suffering a kind of hangover letdown. Hal and Dan slept while I kept awake at the wheel. With the summer wind whipping past the car's windows, I was already thinking of next year, when we might do this again. In the trunk packed in dry ice—proof of our trip's success—was a cooler full of gutted fish.

Later that summer I ran into Dan at Lake Independence. I'd been allowed to take our truck and go to the lake for a swim after evening chores.

There'd been a storm the night before that had detached a section of marsh from across the lake and driven it into the bay in front of Gertz's Resort, where we were swimming. Storms often did that, broke off pieces of bog shoreline from the bigger lakes that became floating islands, cattail rafts, which sailed back and forth across the lakes.

This one was some thirty feet across, in chest-deep water—a challenge.

"I bet I can swim under it," I told Dan.

"You think so?" he said.

I'd bought a pair of swim fins that summer—they were something new on the market, used first by Navy frogmen and Mediterranean skin divers—and they made you feel like a fish in the water.

"Sure," I said. "Go around to the other side and wait for me."

I pulled on my flippers. Took several deep breaths, hyperventilating like the free divers I'd read about, to charge my blood with oxygen. Took a final breath and dove, under the bog, and started swimming.

I stretched out my arms and used my hands as diving planes, keeping close to the bottom, kicking steadily but not too strenuously with my flippers. You conserved your oxygen by not exerting yourself, and tried to imagine that you could hold your breath forever. There were divers, I'd read, who could stay under for four or five minutes. *My* best time, with my head in the stock tank in our barnyard, practicing, was about a minute.

I felt the bog above me pressing closer as I swam under it, closer and closer, until there was hardly any space between its mucky underside and the lake's sandy bottom. I had a mental picture of its size and thought I must be close to the other side of it by now when, bursting to let go of the stale air in my lungs, wanting to *breathe*, I swam into a tangle of muck and roots where the underside of the bog and the lake bottom met, and—of course!—realized that the "floating" bog was in fact grounded. In a sickening, disorienting flash—of panic, claustrophobia, the terror of suffocation—I knew my foolishness and that *this was it*, this was the end of my life. I was going to drown.

And yet, in that dreamlike state one enters in the face of death, detached, still alive, I disentangled myself, turned, and kicked away with my flippers. I felt the bog lift away from the lake bottom and on instinct rolled over and began to pull myself along its underside, grabbing handfuls of the muck to increase my speed until, reaching for another handhold, I felt the edge of the bog and my arm come out of the water. I stood up then, as if resurrected, and breathed—breathed all the sweet, life-giving air on the planet and saw the brilliance of the setting sun and heard the voices of children with such an intense, hallucinatory rush I almost fainted.

"Raaass! Where are you?" Dan was calling from out of sight across the cattails, where I was supposed to have surfaced.

"Here," I said.

WHEN THINGS WERE GOING TOO GOOD

If there was a best time on our farm it was around 1952. That was the year we reached and thereafter held our peak production, milking twenty-eight Holsteins, several with registered pedigrees from purebred bulls through artificial insemination. We no longer had a bull of our own. Instead, when a cow came in heat, Dad called the vet to come with his picture book of bulls and their pedigrees and his vials of their semen Starting that year, we shipped eighteen ten-gallon cans a day from winter to spring and received monthly cheques—December through March—of a thousand dollars or more from Twin City Milk Producers in Minneapolis.

As summer came on, though, and the cows began drying up, the folks earned dwindling milk cheques and ran up bills and we lived on credit. We made hay, cultivated the corn until it was ahead of the weeds, and watched the grain grow from what looked like rows of grass to long, headed-out stems that turned golden in the late summer, signaling harvest time.

After the harvest we plowed. Fall plowing put you a jump ahead of your spring planting. In the spring you prepared the ground for planting, as soon as the snow was gone and the fields were dry enough to get into them, by disking, the disk's angled cylinders breaking up the rounded furrows left by the plow. There was another implement with curved steel blades called a spring tooth that could be used to break up a field and that, in addition, tended to pull up the bigger rocks. The spring tooth made disking easier after a field had been plowed in the spring and the freshly-turned soil, still tightly bound, needed to be raked apart.

You planted oats, and when the ground was warm enough and the leaves traditionally the size of a squirrel's ear, corn. With the oats you might plant a "cover crop" of clover or alfalfa to come up in the stubble after the oats had been harvested. In the spring, clover could be plowed under to provide green manure before planting corn. Alfalfa coming up in the stubble would make a hay field.

Before and sometimes after planting you harrowed, or "dragged," the field, the harrow's spikes breaking up the clods and smoothing the soil.

Until the early fifties we used a grain binder for harvesting the oats, a machine driven by horses in the early days, then by a tractor, that cut swaths through the field and bound the grain into bundles and dropped them onto a fork-like apron that periodically was lowered so that the bundles slid off onto the field. Afterwards you had to shock the grain, walk from pile to pile of the bundles to stand them up for drying. It was scratchy, tedious work, performed out under the sun in the last of the summer's heat, but a lot less strenuous than hefting hay bales. The fun came when you lifted a bundle and uncovered not a snake, which was always startling, but what we called a field mouse—a vole, really. I'd run after it and try to step on it. Once, pretending I was a fox, I "pounced" on a vole and caught it with my hands. Of course, it bit me, and when I stood up yelling, it hung on, dangling from one of my fingers until I shook it off, along with a chunk of that finger.

I could still make a grain shock if I had to. What you did was grab a bundle in each hand and lean them against each other. Then you added another couple of bundles, then another, to make an A-frame of them. Then you put a bundle at each end to bolster the pile, and laid a bundle on top for a "cap." The result was like a thatched little house, and a field of shocked grain looked like a city of little thatched houses.

Then came threshing, or "thrashing," as it was called, which farmers continued to do after the war until they switched to combining—as, maybe a year or two earlier, they'd switched from putting up loose hay to baling. After 1950 you saw junked old threshing machines in farmers' yards or in the weeds on the edges of their fields, along with other rusting, outmoded machinery—one-bottom plows, wheel-driven mowers, dump rakes, once marvels of the industrial revolution but now gone the way of the draft horses that had pulled them.

In 1952 we bought an Allis Chalmers combine, which, pulled by a tractor and driven by the tractor's power takeoff, cut the grain, shook the heads off into a bin, and blew the stems, the straw, into rows. The bin, when full, was emptied into sacks and trucked to our granary. Later the rows of straw were windrowed and baled. It was an incredibly fast, easier way to harvest and made the big, loose pile of straw left each year by the threshing machine, from which bedding had to be forked out for the cows, a thing of the past.

Before the harvest, though, the summer was given to making hay, mostly alfalfa, baling it and putting it up in the barn, and in fact the powerful, legume-y fragrance of mown alfalfa was the essence of summer for me then. To this day the smell of mown hay, which often, in midsummer, is in the air as you drive through the Creston Valley here in British Columbia, sends me instantly back to my boyhood, back to Minnesota, and all those summers when my father and I made hay together.

It must have been that peak year of 1952 when my father and I put up eight thousand bales of alfalfa, by the machine's counter, and was it the following spring that Dad put an ad in the *Farm Journal* or *Successful Farming* and a rancher came from South Dakota or Montana for a big semi-load of our surplus hay at the going rate then for prime alfalfa?

By now, I was learning a little of what my father had to teach me and was starting to know something of the satisfaction of being able to do the work. Especially when I was alone out in a field, on a tractor, away from my father's critical gaze, his blistering words, I was able to dream as I worked. I'd watch the swallows that circled the tractor, reaping the insects kicked up by the machinery, and imagine they were gulls around a fishing boat, *my* boat, off some exotic shore. They were barn swallows, the same cheerful, fork-tailed little creatures that flew in and out of our barn in the summer visiting their mud nests, plastered to the rafters only a few feet above me, as I cleaned the gutters. They cheered me on as I worked.

Then, working together with my father in the fields, in the suffocating hay barn, doing the milking with him, he'd grin at me sometimes. I learned to watch for that occasional, mute acknowledgment.

My mother's last child, named Martin Gerard, was born in February. Nancy, the next youngest, was three. I had three brothers and three sisters now. The oldest of my sisters, Joyce, only fifteen months younger than I, turned sixteen in March of '52 and, taking after our mother, had become a dark-haired beauty. They called her "Frenchie" at school.

The year I was in junior high and Joyce was still in grade school, I took to teasing her. It had something to do, I suppose, with my suppressed recognition of her ripening body. Anyway, she had a spitfire temper and responded so quickly, so satisfyingly, to my smart-ass remarks, I couldn't resist making them. She took a kitchen knife to me once. And once she came at me on the lawn and we grappled until, using a wrestling trick, I hooked my right foot behind her ankles and tripped her. She fell to the ground and lay there howling for half an hour. Her fury made me laugh, though I was ashamed of my meanness.

Then, her first year at Mound and my second, I felt protective of her and we shared confidences. I warned her of guys to stay away from, the studs, the make-out artists, and chaperoned her to one or two of the sock-hops in the school gym, making sure that I, and not some Romeo, took her home. She, in turn, told me a little of what girls talked about when they crowded together into cars in the school parking lot to smoke and gossip during the noon hour. The racier girls, I was excited to learn, talked of "doing it" or "going all the way," including comments on the size of this or that boy's "thing," as Joyce called it, though the experienced girls had the same funny, descriptive words for it that we boys did.

Marcia, three years younger than Joyce, was still in sister school at Loretto. She took after, as I did, our father's German and Celtic side of the family, and developed into her own, fairer kind of beauty. The twins, Mark and Mike, also at SS. Peter & Paul, were nine in 1952, small for their age, and still too young to be of much help on the farm. And there was Nancy.

Little Nancy, age three, was our princess.

There was an extended family gathering in the back yard on our farm that summer, one of several held over the years during lulls in the haying. I happened to document this one with my Brownie Hawkeye.

Doubtless it was a Sunday after-church potluck (boiled wieners on homemade buns, a "hot dish," potato salad, dill and sweet pickles, angel food cake, coffee, Watkins nectar or our raw milk for the kids), with three generations on my mother's side represented: the grandfolks, my grandmother's widowed sister from Hamel, my mother's four younger sisters with their husbands and kids from Minneapolis or its suburbs—and us. After the meal, the tables had been cleared, the jars of pennies brought forth, and there was penny ante poker under the trees until four o'clock, chore time.

I went around that day consciously "covering" the event, taking pictures of my aunts and uncles and cousins, my teenaged sisters, my mother. None of my father, for some reason. Those faded, black-and-white snapshots are history now, frozen images from that distant summer afternoon.

"Things are going too good," Ma kept saying that year. "I'm afraid something bad is going to happen."

But nothing did—not that year, anyway.

My father and I started trading Sundays off. That gave each of us a break from evening chores every other Sunday, though not from the morning milking, which we did together before Mass in Loretto or Hamel. But after that, the day was mine, or his, depending on whose turn it was. The folks usually went visiting on their Sunday off, and I, on *my* Sunday off, might go swimming in Lake Independence or Minnetonka and in the evening to a show in Wayzata or Minneapolis with my buddies, Hal and Dan and Cal Tyson. I drove. My friends didn't have the use of a vehicle, as I did. It was part of the deal I had with my father.

I was *on* one Sunday and late for chores after swimming in Lake Independence. Driving too fast on the gravel road toward our place in our old International truck with its extremely faulty brakes, I came over the hill by the Miller farm to find the road in front of me full of cows—the Millers' herd, being driven from their pasture along Spurzem's Lake to their barn up the road to be milked. I floored the brakes, without slowing down much, then swung desperately into the ditch where the truck's protruding rack snapped off the poles of the Millers' roadside fence, *crack, crack, crack, crack*, until gradually the truck slowed and finally stopped. It was like the clicking down of a Las Vegas roulette wheel and made for a funny story afterwards. At the time, though, all I could do was stare dumbly through the truck window at the Miller boys, who just shook their heads and waved me on.

Again in July that year, after first-crop alfalfa, I was given the family car and, this time, a full week off.

Dan and Hal and I had talked of going to the Black Hills, instead of up north again, then decided the drive, across western Minnesota and clear across South

Dakota, would cost too much for gas, and anyway, we'd spend half our time on the road. So we settled for the tried and true: Detroit Lakes.

There was a bittersweet feeling to this second and what turned out to be the last trip we Three Musketeers would have together. I longed to recapture our experience of the year before, but we were a year older and weren't such boys anymore. Dan and Hal weren't, anyhow. Me, I wanted to *stay* a boy, while the others, obviously, were leaving boyhood behind. They were leaving me behind. I felt betrayed a little.

At Benson's resort once more, Hal talked us into renting Gilly's expensive round-bottomed boat and his outboard motor, rather than one of his old, two-bits-an-hour rowboats. The purist in me was bothered. I missed the lazy creak of the oarlocks, the leisurely pace of a rowboat on the placid lake in the quiet evenings. Instead there was the whine and oily stink of the motor now, if also the effortless speed it gave us to tool across the lake or circumnavigate its shores.

One evening a group of young people came down to the lake and splashed happily in the water. Among them was a lovely, full-breasted girl in a two-piece bathing suit who had Hal and Dan and me simply groaning with desire for her as she swam nearby with her boyfriend. She was as beautiful as Debra Paget, whose picture (cut out of a movie magazine) I carried in my billfold now, as if she were my girl. None of us had a girl then, though Hal and Dan, I think, lost their cherries before they went into the Army a couple of years later.

Unabashedly, we swam naked every day, like the boys on Jackson's Island, and took snapshots of ourselves, including one of me falling back in the water showing full frontal nudity. They were developed at the Wayzata drug store, and probably gave the guy in the darkroom a chuckle.

We dove in the lake for clams one day, like South Sea pearl divers after oysters, using the boat's anchor to sink us to the sandy bottom. We took turns with my diving mask, and filled the boat with clams we found in bunches, like stones sticking up in the sand that glinted in the sunlight some twelve or fifteen feet down in that clear water. Looking up from the bottom you saw the underside of the boat seemingly far above you, and running out of air you clambered to the surface. Afterwards, we saved only a few of the clams to make "clam chowder" (inedible, as even I had to admit), and threw the rest back into the lake.

We were the Three Musketeers! We were Tom Sawyer and Huck Finn and Joe Harper on Jackson's Island! We were South Sea Islanders!

The intensity of my fantasies was almost painful.

The end of August brought the week-long Minnesota State Fair. Fair Week heralded the end of summer and the approach of winter. It was a little sad as well as exciting.

We went to the fair almost every year, making the long drive to the St. Paul fairgrounds to park the car in one of the huge lots, acres away from the fair itself, with the hope we would find it again at the end of the day. Once as a child I got lost in the crowd. The folks' instructions, should you get lost, were to go back to the car and wait. If you couldn't *find* the car, you should look for a cop. Somehow I found our car that day, found it unlocked (who locked their car in those days?), and sat in it and waited. I waited for hours, it seemed, before the folks and my sisters showed up finally. "How long have you been here?" my mother asked, not as concerned as I thought she should be. "We didn't know we'd lost you until just a little while ago."

You waited until it rained to go to the fair, or for some other excuse to take the time off. Some years you missed it because of a polio scare, or too much work to do, or not enough money.

The folks stayed home from the fair that year but let me take their car to go to it with my buddies. We reached the fairground about noon on a hot, sultry day and sweated through the afternoon. Like good country boys, we toured "machinery hill" and the livestock barns, trudging from displays of new farm equipment to stalls of prize cattle and horses, pens of sheep and pigs, chickens, ducks, guinea fowl. At the food stalls we filled up on pop, pronto pups, caramel corn, ice cream, just about every junk food offered for sale, then hit the midway with its thrill rides. The rides, after all that grease and sugar, made me sick.

One fascinating attraction was a daredevil act I think was called barrel racing. I don't mean the racing around actual barrels on horseback you can see in rodeos but riders on motorcycles, who circled the inside of something like a wooden silo, held against the walls by centrifugal force, while you looked down at them from the platform around the outside rim. The structure shook from the weight of the roaring bikes as they raced around and around, climbing the walls so that occasionally a rider and his machine threatened to shoot over the top of the barrel. There was a slanted wire guard around the top to prevent this, and I imagined a rider getting caught in the wire and dropping with his heavy machine to the barrel's floor.

Tough-looking men—and women—performed this feat. What especially fascinated me were the women in their greasy jeans and sleeveless jerseys with tattooed arms or shoulders at a time when nice girls didn't get tattoos. What were they like, these women? One or two of them weren't bad looking, or rather they *were* bad looking, and exciting to think about.

In the evening, feeling reckless and rather worldly, we joined the crowd in front of a girlie show, its garishly painted tent, and listened to the barker's spiel ("Ten be-yootiful girls, eight be-yootiful costumes!" as the parody goes). We got a look at one of the performers when she was brought out to entice us. She was no girl, actually, and even tougher looking than the daredevil motorcyclists.

"Whattaya think?" somebody said.

We bought tickets and swaggered into the tent.

The year before at the fair my father had surprised me, and embarrassed us both, by steering us into a show like this one. Shaking his head as we left the tent afterwards, he'd said abashedly, "Don't tell your ma."

The tent that night was suffocating, crowded with men and boys and even some women, most of them hicks, I thought, like we were. We all sat on boards laid between blocks of wood on the sawdust floor and stared expectantly at the empty stage. Presently a man in a sports shirt came out and sat behind a set of drums. Then another man carrying a saxophone came out, and the show started. The musicians struck up a pounding, raunchy number that brought out a busty woman in a flowing skirt and revealing halter who began dancing energetically to the music. She and the sax player had started a thing—he was crouched and jabbing at her with his instrument, she was meeting it, swaying and bumping and grinding, and we were waiting for her to take her clothes off—when, in a sudden, whooshing gust of wind, the tent lifted away from us and flew off into the night.

For a hushed, guilty moment, nobody moved. Then everybody, performers and spectators alike, as if responding to God's wrath, simply melted away. Left sitting there, my friends and I turned around to see the midway in chaos, lights swaying violently, the air full of flying debris, tents either collapsed or flapping wildly in the wind. "*Tornado!*" somebody hollered. People began streaming toward the gates, moving quietly, purposefully out of the fairgrounds.

It poured rain on the way home. After dropping my friends off, I drove to our place and turned into our yard to find a tree fallen across the driveway. Inside the house I heard how the folks and their visitors, my aunt Clarice and uncle Gene, had been playing cards in the dining room when out of the darkness outside came that train-like roar every Midwesterner fears and listens for during a nighttime storm in the summer, and Uncle Gene was shouting, "Hit the basement!"

Everybody piled down the cellar stairs as the roar passed and it was quiet again.

Only then did my mother remember that she'd left the baby, my kid brother Marty, in his crib upstairs.

The next morning there was unmistakable evidence that a small twister had passed within a hundred feet of the house. It had taken down the box elder tree at the entrance of our short driveway and chewed a path, some fifty feet wide, through the corn field across the driveway. The corn was flattened, twisted, crazily skewed, as if some kind of rotary mower with a dull blade had driven through it. Over at our neighbor's you could see how the slats of his corn crib had been sucked out from under its now sagging roof. Later that day, at the far end of our farm, I found splintered boards from the crib driven like spears into the ground.

Then, during chores that night, my father and I noticed that the cabinet on the wall in the milk house was slightly tilted toward where the storm had passed. And yet no windows had been broken, there was no real damage. The twister had missed us, and my neglected baby brother was okay.

CLASS OF 1953

That fall of 1952 I started my senior year at Orono Consolidated High School. Joyce and I had transferred there the year before from Mound when it was still under construction in a field outside the village of Long Lake, only some five miles away from our farm. My friends Hal and Cal Tyson transferred too, but Dan, along with other Loretto kids, stayed at Mound, and so the Three Musketeers were broken up.

My first year at Orono I discovered Jack London. I'd read *The Call of the Wild* and *White Fang* by then, I'm sure, but knew nothing of London himself. I was only beginning to be interested in the writers of the books I read, and in their writing—how they did it. Now in the school library I found Shannon Garst's *Jack London: Magnet for Adventure,* her biography for juveniles that described his rough boyhood in Oakland, his teenage experiences as a San Francisco Bay oyster pirate, as a Pacific sailor, as a tramp riding the rails across America and Canada—his joining the gold rush to the Klondike, where he struck not gold but rich literary material. Finally the book told of his dedicated, self-taught, ultimately successful struggle to become a writer.

All at once, with the force of revelation, I knew what I wanted to be in life.

The so-called police action in Korea was now admittedly a war. It would be our war, we senior boys thought. We'd all be called up to fight in it after graduation. That gave an edge to our senior year.

In fact, the war ended in July that year, but we all still faced the universal draft, which eventually would call you to serve a two-year stint in the Army. With that hanging over you, some boys, as my friends Hal and Dan did the following year, volunteered for the draft to get it over with. Others, as I would do, gained a temporary exemption by going to college.

I'd started thinking about college after my twelfth-grade English teacher at Orono advised me that a college education would be useful if I meant to be a writer. But how to pay for it? I knew the folks couldn't afford to send me to college, and I was anything but scholarship material. My only alternative was to go into the service first, and to college afterwards on the G.I. Bill.

Volunteering for the Army draft, though, didn't appeal to me. You might serve your two years on some dreary post in the South, never even get out of the States. Besides, the Army didn't seem as romantic as, say, the Navy, or better yet, the Air Force. But an enlistment in either of those services was for four years. Still, in the Navy I might see some of the world, as the recruiting posters promised, and in the Air Force, I might become a pilot.

My great fantasy, before the Korean War ended, had been to become a fighter pilot. I'd followed the war in the Minneapolis *Tribune*, particularly the air war with its reports of engagements between Russian-built MIGs and American Sabrejets. The paper included a daily or weekly tally of the enemy planes shot down by our U.S. aces. Around Christmas of 1952, I watched a television report from Korea about the air war that showed Sabrejets on patrol. Watching those sleek, swept-wing fighters peel off, presumably to engage the enemy, completely hooked me.

Mr. Howard, my favorite teacher at Orono (he taught history, one of my good subjects), had been a bomber pilot during World War II, flying twin-engine B-24s, I think, over North Africa. Once in a while he'd talk about the war and recount some escapade for us. One day he told of buzzing a native's house in Morocco, just for the fun of it, and more or less lifting the flimsy roof off the place. I repeated his story, dwelling on the humor of it, in a feature I wrote for the school newspaper, only to be reprimanded by Mr. Howard, and forced to print a retraction in the next issue. I had to write that I'd taken "liberties" with his story and used it without permission. Thus I learned my first lesson as a journalist about writing for publication and the possibilities of libel.

I'd started writing for the school paper, features and "editorials," as a way of preparing myself to be another Jack London. I'd also, along with only one other boy, enrolled in Typing. It would prove, as it has for other writers—would-be and otherwise—to be the single most useful course I took in high school.

"Creatively," apart from school journalism, I was writing crude parodies of various comic book, radio, and early television heroes, handwritten little stories that I passed around to my classmates for their gratifying laughs. "Tarzan and the Sheik of Shit" was a characteristic title. In a three-pager called "Tarzan and Jane," I indulged in erotic fantasy, and in "The Lone Ranger and Tonto" I had the Masked Man and his Faithful Indian Companion bickering like old marrieds. I wouldn't get serious until the following year when, in college, I was introduced to those literary masters who, other than Mark Twain and Jack London, became my idols. Not surprisingly, for someone of my generation, Ernest Hemingway would top the list.

There was a lot of nervous and sometimes drunken hilarity among us seniors that year as we realized we were on the threshold of adult life—and, for us boys, maybe about to go to war. As seniors we were lords of the school, still under the thumb of the teachers but at the top of the secondary school heap.

Our pinnacle of privilege, the event that signaled our graduate status, was Senior Skip Day. The place to enjoy it was next-door in Wisconsin where, at age eighteen, it was legal to drink 3.2 beer. Not that all of us who set out for Taylors Falls that Senior Skip Day in late May of 1953 were of legal drinking age. My friend Hal, who'd started school at five, was almost a year away from eighteen. I, on the other hand, was almost eighteen and a half. Various other kids were close to, if not of, legal age. In any case, we all got drunk. Had to: going to Wisconsin on Senior Skip Day to drink and carouse was a local high school tradition, an obligatory rite of passage.

A friend from Maple Plain drove, with Hal and me and another farm kid in his car. We had a blowout on the way home. But quickly, showing off our manhood to one another, we jacked the car up and put the spare on. Then we pulled into a garage to have the damaged tire fixed. We were standing around the mechanic at work on our tire, joking about farm chores, when he commented, "So I guess you boys are all farmers, huh?"

"Let's put it this way," I piped, infused with alcoholic wit. "We know what shit is."

There were three deaths pretty close to home that spring and early summer. A Hamel boy some three years older than me, handsome, known as "wild" and a "make-out," developed cancer and died within a couple of months. He had helped us haul hay one evening the summer before. I remembered his dimpled, round-faced good looks, and how I was in awe of him for his reputation with girls, and how he kept winking at me as he responded to my dad's friendly questions, as if he and I were sharing a joke.

That fall, outside the Lake Independence dancehall, he revealed his feelings toward his own father. He'd been kicked out for being drunk and causing a disturbance. A bunch of us stood around listening as he pleaded with the bouncer, slurring his words, to be let back into the hall. "I'll *behave* myself," he said. "You have my word. I'm not *like* my father." His father's word wasn't worth a damn. "*My* word is my bond. It's as good as gold," he told the bouncer, waving his arms wildly for emphasis.

But the bouncer stayed firm. "You're drunk," he said. "Go home and sleep it off."

"Fuck you then!" the boy told him, and lurched to his car and peeled out of the parking lot, spraying dirt, a drunk driver on the loose.

A month or so later the word was out: he was dying of cancer. I saw him only once after that—not that I *wanted* to—on the street in Hamel. The change in him was frightening, and seemed to have happened overnight. In a matter of weeks he'd gone from glowing youth to cadaverous old age, his round face sunken and sallow now, his eyes huge with the enormity, I thought, of what was happening to him. The change in his voice was unnerving. It had degenerated from a pleasing tenor to an eerie bass. "Hi, Buddy," I said. "Hi, Ross," he replied, as if from the grave. By spring he was dead, and the story went around that the last thing he did was sit bolt upright in his bed and cry, "*I'm going!*" That was scary.

Then, later that spring, there was another death. I was cultivating corn one Saturday in a field we'd rented on the old Huar farm when Ma pulled up with my lunch.

"Did you know"—and she named a Loretto boy I'd known in Mound, before being transferred to Orono.

"Well, he's *dead*," she said, as if scolding me. "He was killed last night in a car accident."

He'd been another handsome, wild boy, a teenage alcoholic and a reckless driver who'd had a number of accidents with his own car before this fatal one with his folks' new Studebaker. He ran off the road and into a power pole after maybe falling asleep at the wheel, we heard. Later it was thought he might have had an undiagnosed concussion from his previous accident (he'd totaled his own car the week before) and blacked out. In any case, he was dead. Just like that. And after having left the Hi-Spot Café in Hamel only minutes before to drive home to Loretto. He wasn't drunk that night either, according to the last people to see him alive.

Finally an Orono classmate, a fellow farm boy, died of cancer. I'd wrestled him in gym class that winter, wrestled him to a draw, which I was proud of. But he must have been sick even then; had he been well, I figured, he might have beaten me.

He sat through the graduation ceremony with us on a hot night that June, deathly pale, then rose shakily to accept his diploma. He was so weak by then, so brave, our hearts went out to him, and we all applauded as he tottered off the stage and was taken home. He died a week later.

After graduation I was still undecided about what to do now, which way to go. All I knew was that becoming a jet fighter pilot was unlikely. I was color blind, for one thing, and for another, I hated math—I'd barely passed ninth-grade General Mathematics and had avoided the subject ever since. Dan and Hal now talked of volunteering for the draft. Meanwhile, they'd both found jobs in Minneapolis. I stayed on the farm.

One day Hal stopped by on his way home from work. He found me by the machine shed, helping my father change the oil in the Farmall. Dad was under the tractor, draining the old oil. Following his order, I had six cans of fresh oil lined up on the workbench, punched and ready to pour into the engine. Hal and I stood talking, "shooting the shit," as the saying went, and probably irritating my father, when he said from under the tractor, "All right, *get the oil.*"

I went to the workbench, grabbed a can of oil, and emptied it into the crankcase. I was pouring another can into it, still talking to Hal, when my father yelled, "*Hold it*, for Christ sake!" He rolled out from under the tractor, his face dripping with oil. "What in *hell* were you thinking of? I didn't have the plug in yet! I meant for you to get the oil *ready*, not pour it in!"

It seemed nothing was going to change. My father would bawl me out for the rest of my life, and I would take it. He rolled under the tractor again, cursing about my failure to do anything right. Hal put his hand to his mouth to keep from laughing.

I walked away, and Hal followed. Around the corner of the pump house, I grinned unhappily at my friend as we listened to my father ranting on under the tractor.

That summer Dan and Hal wouldn't be free to go up north with me again because of their jobs. The thought of going up north by myself was depressing, and I was resigned to giving up what had become my annual vacation when Dad said, "Hell, *I'll* go with you."

But where to go? We decided—I'm not sure who got the idea first—on a canoe trip in the Boundary Waters of Minnesota, the so-called Roadless Area along the Canadian border. It promised what today would be called a father-and-son bonding, but I wondered how much fun it would be. I'd worked for my father most of my young life, but I was still shy with him. He was shy with me. What would we talk about, away from the farm?

"Could be the trip of a lifetime," Dad said.

"Yes," my mother told us. "It'll give you two a chance to get to know each other."

CANOE COUNTRY

The picture postcard sent to my mother, in my father's sweeping, legible hand and marked with X's for kisses (the picture is of an empty highway curving through "the forests of the North"), is postmarked Ely, Minn., July 13, 1953:

> *Dear Ruth*
> *Got here at 6 o'clock. Everything O.K.*
> *Miss farm. Mervin. X X X*
> *P.S. Tell the boys that*
> *milk to use two pads*
> *in Milk strainer.*

The P. S. referred to a couple of the Miller boys Dad had hired to milk our cows while we were gone. July 13 was a Monday; we'd reached Ely, in the northeastern corner of Minnesota, on Sunday evening after a three hundred-mile drive, and put up in a motel. Monday morning we presented ourselves to Bill Rom at Canoe Country Outfitters (we'd been in touch by mail), who set us up for our five-day trip into the wilderness. We were given a sheet of camping instructions, detailed maps, a compass, canoe paddles, and three duffle bags containing a tent, sleeping bags, an air mattress, cooking gear, waterproof matches, and dried and canned "grub." We loaded the car. "Have a good trip," Rom told us.

We drove to where the highway ended at the little town of Winton, then past it on a gravel road to a turn-off that took us to the shore of Moose Lake, our starting point. There we found a rack of aluminum canoes. We picked out a two-man fifteen footer and carried it to the water. Then we transferred our gear from the car to the canoe, locked the car, and pushed off into the lake. There wasn't another soul in sight.

We had Skippy, our little Irish setter-cross, with us, my father's idea. (We'd sneaked him into the motel with us the night before.) He'd keep us company and warn us of bears, Dad thought. We were both a little afraid of bears, though the outfitter had told us how not to attract them: hang your food and garbage from a tree, don't keep food in the tent, *don't leave any garbage.*

Skippy took to the canoe without fuss: jumped in and settled himself among the packs between the seats and lay quietly as we paddled. The country was all clean water and granite rock and unbroken forest. The water sparkled in the sun. You could drink it.

We were on our first portage, carrying the overturned canoe on our shoulders, just coming into sight of the lake in front of us, when a thunderstorm hit with a rush of wind and then lightning and thunder and a chilly downpour that hammered on the bottom of the canoe as we crouched under it on the trail. Skippy whined and trembled between us.

"What if it rains like this the whole trip?" I wondered.

"Let's hope it don't," Dad said.

Then the rain stopped and the sun came out and the trees glistened. We stood up and walked down to the lake, the water sparkling again, our farmer's high work shoes slipping on the rocks or sinking into the humus. We dropped the canoe on the lakeshore, then went back for the three duffles, including the food bag, which we hadn't forgotten to hang from a tree.

We were ready to push off again when we sighted another canoe. It appeared as a dot far out in the lake and advanced slowly toward us. You could see paddles flashing and then the canoe itself and then its occupants, a young couple. They turned out to be newlyweds, heading back after only two days. The pretty young wife, splotchy with mosquito bites and complaining of blistered hands and sore muscles, was outspokenly angry with her husband for making her rough it on their honeymoon. The young husband, smiling ruefully, told us they would finish their honeymoon at some picturesque resort overlooking Lake Superior. Dad and I had a laugh about them after they left us. They were almost the only people we saw that week, and I could pretend we were seventeenth-century French *voyageurs* or *coureur de bois*, the first white men to see this country.

Toward evening after a long day of paddling and portaging, we came out on Knife Lake, along the Minnesota-Ontario border. The lake's forested shores stretched into the distance. Away off, in the middle of the lake, was a good-sized island. We made for it, thinking to camp there. Drawing near, we saw that what had appeared to be a single island was actually three of them, one relatively large and two smaller islands, all connected by foot bridges.

"Hey, looks like somebody lives here," Dad said. "Let's see who it is."

We paddled under one of the bridges and found a landing on the main island. An overturned canoe was there, and a path leading up a steep slope to the island's wooded top.

We walked under the trees and found a log cabin overlooking the lake. A couple of smaller cabins stood nearby. "Nobody's home," my father said. But then, snooping around to the back of the place, we came upon a strapping, youngish-looking woman in a one-piece bathing suit, standing on a diving board wedged into a cliff jutting out over the water.

"Oh! Hello!" she said. "I'll be right with you."

She dove into the lake, then swam in a strong crawl to the granite shelf below us, climbed the rocky slope to where we stood, dumbfounded, and grabbed the towel she'd left hanging on a tree. She had an athlete's firm body and short curly hair that she rubbed hard with her towel. Dad and I stood shyly watching her, but she seemed perfectly at ease.

"You want some root beer?" she asked. "I make it myself."

She had a cooler on the roofed front porch of the cabin, filled with ice water and bottles of her delicious, homemade root beer.

We sat on the porch, drank root beer, and introduced ourselves. Then: "C'mon," she said. "I'll show you around."

We followed her across the bridge to the nearer of the small islands, where she showed us her ice house. It was a windowless log building standing in the deep shade of the trees. She opened the door and we stepped onto a wet stone walkway inside between wooden cribs filled with blocks of ice packed in sawdust. The woman cut the ice on the frozen lake in the winter with an old gas-powered saw. "Keeps all summer," she said.

Then we crossed to the farther island, which was covered with enormous old conifers, virgin Norway and white pine, the woman said. She showed us a tree with a scarred-over slash in its bark that wasn't a timber cruiser's mark, she said, but a trail blaze left from the days of the fur trade, around 1800.

"Is that so?" my father said. We both stared at it. I thought of the nameless voyageur, all those years ago, swinging his hatchet to make that mark. I thought of the handprint of a Pleistocene Age hunter on a cave wall I'd seen pictured in the *National Geographic*. This blaze, that handprint, was human evidence from so far in the past that all you could do was stare at it.

Before we left her, the woman told us of an island around the next bend in the lake where we could camp. I think she may have invited us for supper, but my father wanted to set up camp before dark and the nightly onslaught of mosquitoes.

"Come back after your supper, then, and we'll visit," she said.

"We might just do that," said my father.

We found the little island she'd mentioned. It was just a humped piece of rock sticking out of the lake with a few scraggly trees on it, very close to the south shore. Evidently it was a regular campsite because we found a rock-lined fire pit and pieces of weathered rope on the trees where tents had been pitched or clotheslines strung.

As we were setting up camp, we saw a beaver. I saw it first, a V in the water and then the blunt head of an animal almost the size of our dog, swimming toward the island. "Look, Dad!" I called. At the sound of my voice the beaver flipped under the surface with an identifying loud smack of its tail against the water.

That's when we saw its house, a huge pile of mud and sticks poking out of the water, just down the shore from the island.

We made a supper of wieners and canned beans or canned spaghetti, something easy from our food pack, and washed it down with Kool-Aid made with lake water. Then we scoured our tin plates with sand in the lake, bagged our garbage, and hung both garbage and food duffles in trees well out of reach, we hoped, of bears or wolverines. Then we got into the canoe and paddled back to the woman's place.

The inside of her cabin was one long room with a wood stove in the middle of it, a trestle table at the far end, and shelves all along the walls. There was a ladder at one end, leading up to her sleeping loft, and a wire stretched under the cabin's low ceiling with a Coleman lamp hanging from it that could be moved along the wire. It reminded me of the setup in my grandpa's barn when he was farming. The woman positioned the lamp above the table and we sat at it with the Coleman hissing over our heads and drank coffee or perhaps more of her root beer. She and my father talked and talked.

Her name was Dorothy Molter, and she was a nurse, originally from Chicago. She'd been living up here permanently since 1934, and alone for the last five years. She'd acquired this place after working for the original owner and finally nursing him until his death in 1948.

"Don't you ever get lonesome?" my father asked.

"Nope", she said. She had plenty of visitors in the summer, people on canoe trips like us, people who came to stay here at her resort, called Isle of Pines. She had an outboard motor for her canoe, and in summer brought in supplies from Winton or Ely; in winter she could always snowshoe out across the frozen lakes, and she even went back to Chicago, once in a while.

"You do the portages by yourself?" Dad asked. "With a motor?"

"Takes a couple of trips, maybe three," she said. "Motor's pretty small. I sling it over my back, transfer the canoe and my supplies in a couple of carries. I got it down to a system."

"I bet you have," said Dad.

She was a little tomboyish and too old for me, but I could see my father found her attractive and liked talking with her (he talked easily with strangers, was always curious about them: what was *their* story, how did they make a living?). She kept smiling broadly as she answered his many personal questions, then questioned him in turn. Was there a mutual attraction? Maybe. In any case, my father was no womanizer. He was much too shy, too inhibited, too married to my mother to even flirt with another woman. I doubt he knew how attractive he might be to women other than my mother with his sharply handsome features and black hair without a hint of gray in it yet. He was forty-two that summer.

Dorothy Molter was forty-six then. We read a story about her in the Minneapolis *Star* or *Tribune* after we got home. The paper called her the "Wilderness Nurse." Later she became famous in the north country as the "Root Beer Lady," and remained on her islands until her death in 1986, after which her cabins, derelict by then, were dismantled, transported to Ely, and restored to form the Dorothy Molter Museum.

It was getting dark when we left her place and paddled back to our campsite. By the light of our flashlight we crawled into our tent, tied the flaps down, and zipped up the mosquito netting. Then, ignorant as most people were then of the health hazards, we sprayed the inside of the tent with DDT, snuggled into our sleeping bags, and pretty soon my father was snoring.

I was awakened at first light the next morning by the eerie cry of a loon, a wild sound I'd first heard on my Detroit Lakes camping trips. Listening for it again, I heard what sounded like a train passing to the north of us. Looking at our maps later that morning, I found no railroad indicated, on either side of the border, only lakes and swamps and campsites and elevations. That and three or four little squares marked "resorts," where the road from Winton branched out at the head of Moose Lake like the delta of a stream. What I'd heard, I finally realized, was the rush of the wind through all that wilderness of trees surrounding us.

When I walked to the lee shore of our little island that morning for water, I saw how it dropped steeply into the channel between the island and the near shore. You could see the bottom of the channel clearly, maybe twenty feet down. Before breakfast (flapjacks and bacon, whipped up by my father, who always claimed to my mother that he didn't know how to cook), I dove off the island with my mask and flippers, down past the warm surface layer of the water to where it was heart-stoppingly cold and the jagged rocks along the bottom were furry with algae. I swam into a shadowy alcove under the sheer underwater face of the island and flushed a fish, a big walleye or muskie, which whipped away, then halted, suspended in the water, to look back at me. I imagined myself in some tropic sea, encountering a shark or a barracuda.

After breakfast Dad said, "Hell, let's go back to the outfitter's and get a motor like that woman uses. This is supposed to be a vacation. I don't know about you, but *I* didn't come up here to paddle my ass off."

Outboard motors are banned now in the Boundary Waters, but then they were allowed (as were the snowmobiles in the sixties and seventies that brought winter visitors to Isle of Pines). Still, an outboard seemed a violation in that pristine region, and I pretended to myself to be disappointed in my father. In fact, I was as ready as he was to stop paddling *my* ass off.

We had a long paddle ahead of us, though, to get back to our car. We struggled most of that day against a strong headwind and high waves on the lakes that had us hugging the shorelines, taking a roundabout way. But we gained some time by running the modest rapids in the Knife River, which we'd had to portage around on the way in. We made it, at last, to the south shore of Moose Lake and the parking lot where we'd left our car. Drove to the outfitter's in Ely. Got a small outboard motor, then drove back to the parking lot. Picked out a flat-stern canoe and put it in the water at the end of the dock. Then, while trying to attach the motor to the canoe, my father dropped it into the lake. Down it sank, through the clear water, to the bottom.

"Christ. Whatta we do now?" Dad said.

"I can dive for it," I told him.

"You can?"

Naked, in my mask and flippers, I slipped off the dock and kicked down to where the motor lay, reflecting the sunlight, on the rocky bottom some fifteen feet down. Grasping the motor, I kicked for the surface while my father's grinning, wavering face peered over the dock at me through the water.

The motor lifted easily enough until I reached the surface but then was heavy as a rock. My father just managed to get hold of it and lift it onto the dock. I pulled myself onto the dock and had him look at me as if seeing me for the first time. "I never would have thought you could do that!"

We dried out the motor and got it started; set off down the lake with Dad at the rudder and the little motor purring, the canoe cutting through the water at a good clip. I looked back to see a big grin on my father's face. We headed north into Canada.

We had another couple of days. The mornings were foggy, chilly, until the sun broke through. Then it was glorious. We didn't see any animals other than the beaver that one evening: no bears, regretfully, no moose. Skippy, his nose up, would indicate he'd caught some scent in the air, and once in a while he barked. That served to scare away, I suppose, whatever critters we might have seen.

The faded pictures in my photo album of the trip include several taken from a high point that we must have climbed for the view. They show vistas of sky and water, forested hills, and round, forested islands in a large lake. The lake is empty.

Floating down a river we passed over sunken log jams, huge jumbled piles of timber cut fifty or sixty years before, around the turn of the twentieth century, and now saturated, yet preserved, on the bottom. In the crystal water we floated above their clutter as if on air.

"Christ," my father said. "What a waste. There must be a way to salvage some of that wood."

In the tent at night, we talked. We talked into the wee hours, more than we ever had before or would again. My father would die before we got another chance to commune with each other as we did on that trip.

What did we talk about? We didn't confide in each other, as my friends and I did. I could never do that with my father, nor he with me. And yet those nights as we lay in a tent together in the canoe country wilds—in the guarded way of men and the extremely guarded way of my father and me—we did get to know each other a little, as my mother had hoped we would. I saw my father differently after that, saw that this self-made, often angry man, who used to frighten me and still growled at me, still humiliated me at times, had a soft side, one he only showed to my mother and occasionally to my sisters—and to *me*, sometimes, now that I thought of it.

Those nights in our tent he picked my brain about what I knew, what I'd read. It was some kind of vindication of all those times when I was supposed to be dumping milk for him and *paying attention* and instead had my nose in a book or sat listening to the radio in the barn. That used to exasperate the hell out of him. Now he wanted to know what I'd learned in those stolen moments.

Turned out he was fascinated by history, as I was, but had read little since dropping out of high school except farm magazines and the daily newspaper. What he wanted to know now was how "the average guy" lived in olden times, what kind of work people did in, say, ancient Egypt or during the Roman Empire, what they ate, what they wore, the kind of houses they lived in, the look of their towns, the look of the country. What was it like when *this* country was still wild and a man could live like the Indians, sometimes *with* the Indians, and there were woodsmen and mountain men, men like Simon Kenton and Daniel Boone and Jim Bridger and Kit Carson (men my father had never heard of until I told him about them), men who had loved the wild country but ironically helped to tame it, to fill it with settlers, when they took to guiding wagon trains or scouting for the Army. I think my father and I both would have liked to have lived then.

I was at once flattered and unsettled by his respectful, yes, *respectful* listening to me (*he* was the one paying attention now, taking an interest), and by his apparent belief that I knew what I was talking about. Of course my reading, as I discovered when I got to college, had been scattered and undisciplined, and my "sources" not always reliable. My "knowledge," therefore, was rather suspect and full of holes. But never mind. When I found myself stumped by one of my father's questions, I extrapolated. I bullshitted, in other words.

We talked about my leaving.

"You sure I can't convince you to stay on the farm?" my father asked. And not for the first time, he added, "You stick with me, you know, and the farm'll be yours someday."

I knew that. He was offering me his kingdom, and I was rejecting it. I'd decided, at last, to enlist in either the Air Force or the Navy (I didn't know which yet) after the fall harvest that year, though just the thought of it made me homesick already, filled me with guilt. I was my father's oldest son, after all, and my twin brothers weren't old enough yet to do a man's work. The temptation was to stay. But I had to go.

Our last night out, camped on an island in a lonely lake, our solitude was broken by the noisy arrival of a party of men, who set up camp on the mainland across from us. Skippy barked at them as if they'd invaded our property, and we yelled hellos over the water. They came in three canoes, one of them, apparently, loaded with beer. We watched them fill a big fishnet with canned beer and lower it with a makeshift winch into the lake. They kept us awake much of that night with their drunken talk and laughter. Periodically we'd hear the creak of the winch as they hoisted more beer out of the water.

We started back the next morning. It was late in the day when we reached Ely, and we decided to wait until morning before driving home. We registered at the same motel in which we'd stayed at the start of that week (sneaking Skippy, as before, into our unit), and spent the evening in Ely, eating in a restaurant and going to a movie afterwards.

Early the next morning we drove southeast toward Lake Superior. The day turned hot, and coming up over the ridge above the lake we saw the horizon, the whole wide world in front of us, suddenly fill with water that stretched away to meet the sky. Then felt a delicious coolness, like air-conditioning, as we descended to the shore of that cold inland sea.

Driving southwest now along the shore, we stopped at Split Rock Lighthouse and I took a picture of it. We passed Gooseberry Falls, the town of Two Harbors, and reached Duluth. My father ran away to Duluth when he was fifteen, he told me, intending to sign on as a lake sailor. He hitchhiked, and got a ride from a man who by the time they reached Duluth had talked him into returning home. The man took him to the Greyhound depot and bought him a ticket to Minneapolis, then made sure he got on the bus.

We had a blowout somewhere south of Duluth. I was driving, my father asleep beside me. There was a pop, and I hung onto the wheel until I could pull over beside the road. Startled awake, Dad said, "What happened?" Then: "Good driving."

We put the spare on and he took the wheel. The blown tire was shredded, but we didn't stop to buy a replacement. Just kept driving, hoping to make it home without getting another flat. We reached St. Paul, crossed the Mendota Bridge into Minneapolis, drove on through. Then we were west of the Cities, almost home.

As we neared the farm, I could feel my father tensing up. He started talking about all the work we had to do now, second-crop alfalfa, combining, silo filling. I felt myself tighten. Our vacation was over.

"Tell us about your trip!" my mother said brightly after we'd pulled into the yard and walked up to the house to be met on the porch by her and my sisters, the twins and little Nancy. One-year-old Marty was inside, taking his nap.

"Ask Ross here. He could've stayed up there all summer, I think."

The trip of a lifetime: that's what my father had hoped it would be, and that's what it was. But I had mixed feelings about it later because of what happened later. I'm sure my father did too. There's an old saying: After every joy, there comes a sorrow.

The accident

Friday, July 24,1953. 9:45 a.m.

My father and I had been back just a week from our canoe trip. We'd finished the milking and let the cows out to pasture, then walked up to the house for breakfast. After breakfast I'd returned to the barn to shovel out the gutters while my father went to the machine shed to grease the mower and hitch it to the Farmall. He was sending me out to cut hay after I'd cleaned the barn. We were starting second-crop alfalfa.

The night before the ten-year-old twins, Mark and Mike, four-year-old Nancy and fourteen-year-old Marcia, had slept out on the wooded ridge across the heifer pasture. It was a little slice of wilderness within shouting distance of the farm, perfect for first-time campers.

My mother had saved breakfast for them. They showed up about 9:30 after a fitful, mosquito-plagued night on the hard ground, first Marcia with Nancy and then the sleepy-eyed twins. *She came into the house that morning carrying her wet panties. She was still wetting at night. Maybe there was something wrong with her bladder.*

After chores the night before, I'd walked across the pasture and into the trees to see how the rookie campers were doing. Earlier I'd helped them pitch my pup tent and make a fire so they could roast wieners and marshmallows, and now they were bedded down in the tent, Nancy snug between the twins. I knelt and hugged her, said, "Good night, little one." She looked at me then, her big brother, in such a curious, pointed way, as if wondering why I had done that.

That is my last image of her, alive.

The twins had eaten already. I put dry pants on her and had cereal for her on the table. But then she heard the tractor start up and ran out with the twins before she'd eaten anything.

I'd finished cleaning the barn when I heard the old F-20 start up and move across the yard toward the house. Then the chucking sound of the mower's cross-cutting blades. There was an island of weeds and long grass around the light pole in the yard, between the house and the barn, and I knew my father was cutting there with the mower, doing something useful while he waited for me to come up from the barn and take the tractor from him. That gave me a few minutes in which to play.

I was eighteen years old! And there I was, kneeling on the cement floor of the barn, lost in fantasy, *sword fighting* with a couple of screwdrivers, when I heard the tractor shut off and then, after a pause, such an anguished howl and then such a chorus of howls from up by the house that the sound will haunt my memory forever. It brought me running out of the barn to see Joyce and my mother in a kind of crazy dance on the front lawn, passing something back and forth that they took turns hugging, and I thought, *It's Nancy. She got caught in the mower.*

I ran up to the lawn as Marcia stood stricken with little Marty in her arms staring wide-eyed around him, and the twins stood dumbfounded, rocking from side to side as they did when upset, while my father, with Joyce and my mother circling frantically around him taking turns carrying Nancy's limp body, simply stood there. Finally he said, "Put her *down*, for chrissake."

Joyce, I think, was holding Nancy then, and she laid her down on the grass. I saw something gray, gutlike, sticking out of her mouth that a fly had landed on. I bent down and flicked it off.

When I straightened it was to look into my father's suffering eyes. "She's gone," he said with a terrible finality. "*I didn't see her.*"

I looked down at my little sister on the ground, at her child's body with the life crushed out of it, thinking, *This is real. This really happened.* And with that detachment that writers and even would-be writers have, tearless but with a sudden, sick headache, I thought, *I'll have to write about this someday.*

What had happened was one of those tragic farm accidents you read about in the back pages of city newspapers:

Father's Tractor Kills Girl, 4

Crushed by a tractor driven by her father, Nancy Klatte, 4, was killed instantly today near the Klatte farm about four miles west of Hamel in rural Hennepin county. The father, Mervin R. Klatte, was cutting weeds on his farm along county road 9. The child was playing nearby. She suddenly darted in front of the vehicle and was caught by one of the tractor wheels.

The Minneapolis paper got it wrong. Nancy hadn't run in front of the tractor but behind it. The details were these: Dad, cutting weeds around the light pole, looked up to see Nancy and the twins standing on the edge of the lawn, watching him. He waved to them, waved especially to Nancy. He'd let her ride on the tractor with him occasionally, putting her on his lap behind the wheel, and maybe she took his wave now for an invitation. In any case, Mark (Mike had turned away and was walking back to the house) saw it happen: saw his little sister suddenly run down to the tractor as Dad, turned back to his work, drove the mower up to the pole, stopped, then shifted the tractor into reverse. She was behind it, struggling to climb onto the tractor's drawbar when, without looking (because who would have thought?), my father backed up. Nancy fell, fell behind one of the great rear wheels of the tractor, and it rolled over her.

Somebody must have covered the little body. Somebody must have called the sheriff. We drifted into the house, where my mother went to the wall phone in the kitchen and hysterically started calling everybody she could think of. *We had an accident! Nancy's dead!* Afterwards we sat silently around the table, staring at one another.

Until, glaring at me as if with hatred, my mother said, *"For God's sake, take off those stinking overshoes!"* Which is when I realized I'd walked into the house still wearing them, covered with shit after cleaning the barn.

Then people starting arriving: the sheriff and probably the county coroner, neighbors, our parish priest.

Liz Herman, notorious for listening in on the party line and whom my mother thought of as an old busybody, walked into the house, rolled up her sleeves, and took over. After that, my mother could only speak well of her.

The grandfolks needed telling. I suppose Ma had called their number, but nobody had answered. "Dad's working," she said, "and Mother's probably out in her garden. Somebody better go to Hamel and tell her."

Father Michael looked at me. "Ross and I can do it."

We drove in his car to Hamel and found my grandmother in the house, making lunch for herself. She seemed to know why we had come.

Father Michael said, "There's been an accident on the Klatte farm. Little Nancy . . ."

"She's dead, isn't she," Grandma said dully. She had a stroke soon after that, and overnight, it seemed, she was old.

Back at our farm the house was full of people, some making themselves useful, others just sitting around—there for whatever support they might give to the grief-stricken. Liz Herman was cooking food for everybody.

The folks and my sisters sat in a stupor. Somebody held little Martin. The twins were somewhere, traumatized.

That afternoon my uncle George, Dad's older brother, arrived from his farm outside Kimball to help with chores. He noticed our cows were out. They'd trampled down a section of barbed wire fence and were in the cornfield next to the pasture.

"C'mon," Uncle George said. "You and I better get those cows in before they bloat on that green corn."

Still with a numbing headache and finding it hard to move, I helped drive the cows back to their pasture. Then with steel posts, the post driver, wire clips and the block-and-tackle fence stretcher from the machine shed, Uncle George and I fixed the fence. And after the supper prepared by Liz Herman, we did the milking.

The wake lasted three days, three long, mournful days during which Nancy's body was laid out in our parlor. Uncle George had gone home, but Aunt Betty and Uncle Gerry (no longer farming: owners now of the hardware store in Loretto) stayed with us for the time. Uncle Gerry helped with the milking. My father, after a stunned couple of days, got up out of his chair and went back to farming; and, after the funeral, got back on the tractor that had killed our Nancy.

He had the ability, it seemed, to internalize his sorrow and what must have been a desperate longing for things to have been different—his no doubt profound wish that he'd looked behind him before backing up the tractor. Was there guilt because of his "carelessness," which nobody accused him of, nobody blamed him for? Accidents happened on a farm. He himself, with half his foot cut off, was proof of that. But was he ever able to forgive himself for what we had to accept as blind fate, or bad luck, or God's will—or just our human failure to be forever on guard for whatever bad might happen? Or did he carry the guilt for Nancy's death to his grave?

Those mornings during the wake I'd walk over to the open coffin and look down at my sister's swollen, dead face (the morticians hadn't quite succeeded in disguising the crushing effect of the tractor wheel). "She's in heaven now," we told ourselves. I believed it superficially, but deep down I'd begun to wonder if maybe things didn't just stop for you when you died. Maybe you just went back to where you came from, into nothing, as if you'd never lived, never been born.

I wanted to cry but couldn't. I can't yet, as I write this.

The funeral was held on the Monday following the accident. It began at nine a.m. in our house, and continued, at 9:30, in the Loretto church. Father Michael spoke the eulogy: something about Nancy's innocence and how she was an angel now. My twin brothers served as pall bearers, along with four other boys from the parish. They carried the coffin out to the hearse, and then the funeral cavalcade drove the mile or so outside the village to the cemetery, on a hill overlooking fields that lay verdant in the July sun. It was a hot, sultry day, not a breeze stirring, and I stood sweating in my high school graduation suit, surrounded by family and friends and neighbors, over the freshly-dug grave. Father Michael said more words, then shook holy water over the coffin as it was lowered into the ground. Then we turned away. Later, discreetly, the grave would be filled in by the men who'd dug it. They happened to be my friend Hal and his uncle, who was the official gravedigger for the parish.

A couple of weeks later, after church one Sunday, we went to place flowers on Nancy's grave. The folks had bought a small, flatly-placed marker, one they could afford and the kind that cemetery attendants like because they can mow over it. Its simple ornament, chiseled into the stone, is a winged little face, the face of an angel, under which are the words:

Daughter
Nancy Louise Klatte
1949 - 1953

In the days and weeks following the accident, we had hay to make and cows to milk, grain to combine and the silo to fill, corn to pick and the fall plowing. Work can take your mind off things, but they come back to you when the work is done.

At night, my folks grieved together in their bed. I heard them from upstairs through the floor register in my bedroom, above the open door of their bedroom; I heard my mother's weeping and my father's helpless attempts to comfort her. "It was an *accident*, Ruthie!" he told her, over and over. He must have told himself that, over and over, perhaps for the rest of his life.

My mother found a private way to relieve her feelings.

I used to go in the junk room upstairs and close the door. Then I'd bawl my eyes out where you kids couldn't hear me.

Though *I* never cried, I was moved to change. I changed by paying attention, finally, by working as my father wanted me to, by rejecting fantasy and facing reality at last, by stopping what I knew to be childish and even shameful. So I put away, like toys I'd outgrown, Kianga and the imaginary, alternative world I'd created for him, though he's never quite left me. To this day I can switch into his view of things, shrink down to his size and see the world enlarge, the terrain around me, the mountains and the trees and the large lake I live beside become, as it were, another world within this world. It's another way of looking. But I don't indulge in it anymore.

My sisters changed too by growing closer, by becoming best friends. At bedtime, they visited each other's rooms upstairs, like neighbors visiting each other's houses, putting off, it seemed, being alone in their rooms. The twins too, in their bunk bed in the room at the end of the hall upstairs, chattered into the night.

This was after Mike had something like a nervous breakdown. He cried a lot, grew afraid of dying, couldn't sleep.

Marcia hallucinated for a time, felt herself growing larger, or smaller, felt threatened in some way.

Joyce was like our dad, I think. She pushed herself, pushed away her sadness, helped our mother, didn't talk about it.

One-year-old Marty must have been the least affected. He still slept in his crib in the folks' bedroom, wasn't awakened by their sometimes audible sorrowing. My mother lavished attention on him now, spoiling him a little.

As for me, I never properly grieved, I suppose. Instead, I have a nightmare occasionally in which I am overcome by the dreadful, remorseful certainty that I have killed someone—a child, revealingly enough—and hidden the body, buried it under the house, under the floor, like some haunted character in an Edgar Allan Poe story. In my dream I am filled with the horror of it, filled with a longing for it not to have happened, and when I wake—usually I'm jolted awake by the dream itself—I have the awful, lingering sense that the dream is real, a repressed memory, crying, *crying* to be released.

Leaving the farm

Instead of going into the service, I started college that fall at St. Thomas College in St. Paul, an expensive Catholic boys' school affiliated with Notre Dame University in Indiana. I sweated through the entrance exams and somehow passed, enrolled in the Air Force R.O.T.C. (Reserve Officers Training Corps) unit at the college, and declared that history would be my major. St. Thomas was the place for good Catholic boys, Father Michael had advised my folks, rather than the much cheaper University of Minnesota, which might turn you into an atheist or a communist—or both.

The chance to go to college before, rather than after, the service, came when my father said, "Stick around. For a couple more years anyhow. By then the twins should be old enough to take your place. I'll send you to college, if that's what you want. That'll keep you out of the draft."

The tuition per semester at St. Thomas was $750, plus incidental fees: $825. Double that and the figure was $1,650 for the year. Plus books. So where would we get the money? We got most of it, as it turned out, from Nancy's life insurance, a $1,000 policy taken out after Nancy's birth through Ma's insurance salesman cousin.

I car-pooled to St. Thomas that year with three Hamel boys. Two of them, farm boys like me, would eventually become my brothers-in-law, though neither was going with either of my sisters at the time. The third Hamel boy, the owner and driver of the car, was the banker's son, and in fact was the boy who'd broken my horse Buck for me. He was the oldest of us, a senior at St. Thomas that year; the other two were sophomores. We each paid him fifty cents a ride (five dollars a week) for the trip from Hamel to Minneapolis and then across the Mississippi to St. Paul and along the bluffs to St. Thomas's yellow stone bastion above the river.

Weekday mornings then, too early to clean the barn before I left, my mother drove me to Herb's Garage, outside Hamel, where I caught my ride. Back in the late afternoon from St. Thomas, I'd be let off at my grandfolks' in Hamel, where my mother would pick me up. Toward the end of that school year, I'd walk into the grandfolks' house to find them watching the Army-McCarthy Hearings on television. I'd sit down with them to delight in the Army's sane counsel, Joseph Welch, putting the seemingly crazy junior Republican Senator from Wisconsin, Joseph McCarthy, in his place. McCarthy, with his nasal insinuations about communists or communist sympathizers in every walk of American life, his nasty character, made him somebody you loved to hate. I hated him without knowing, really, what the Hearings were all about.

I think now of my mother's sacrifice, getting up earlier than usual to drive me to Hamel to catch my ride to college. I think of my father, excusing me from morning chores so I could catch my ride. I think of him letting me leave the barn before the evening milking was done so I could go up to the house and study. I think of my sister Joyce, who helped me review before my tests.

Looking back, I see that my family's help to get me through my first year of college after Nancy's death was an act of faith, an act of hope, an act of love, though we all knew that my starting college was the start of my leaving.

But then all of us, hardly a year and a half after the accident, would leave the farm.

At St. Thomas, I was introduced to the work of Ernest Hemingway when one day, in English Composition class, our stern instructor, Mr. Scanlan, read to us from Chapter X of *The Sun Also Rises*—its terse, flowing, lovely description of driving over the Pyrenees into Spain in the 1920s. Then he read from Chapter XII, the novel's jaunty, clipped dialogue between Jake and his friend Bill, its studied run of simple declarative sentences suggesting the beauty of the country and the joy of trout fishing. Mr. Scanlan looked up afterwards, his eyes glistening, to see if he'd imparted to us some of his admiration of Hemingway's writing.

His assigned reading was my introduction to other notable American writers: Stephen Crane and Sherwood Anderson (two of Hemingway's influences, I learned), F. Scott Fitzgerald, Thomas Wolfe, William Faulkner. By way of contrast or perhaps as a test of our sensibilities, we were assigned to read an old-fashioned, romantic story by John Galsworthy, called "The Apple Tree." After asking for our reactions (one somewhat effeminate boy said he liked the story; I did too, though I kept quiet), the teacher made a sour face, then told us Galsworthy's sentimentality made him want to vomit.

I wrote a paper for him, "The Plains Indian and the Buffalo," complete with footnotes and a bibliography, as per his instructions, and got a B+ for it. I got a B in his course, both semesters, managed a couple of A's in history, my intended major, but averaged only C+ for the year. I'd have to raise that to at least a B, my advisor told me, if I hoped to graduate from St. Thomas.

That summer, sitting on the wooded ridge across the heifer pasture with a notebook in my lap, I wrote my first attempt at a short story. I called it "Reprieve." It was based on my camping trips up north with my buddies. I can see it now as an amateur, embarrassing pastiche of Hemingway and Faulkner, with an opening sentence that mimics Stephen Crane's in *The Red Badge of Courage*, but it expressed my longing for escape—escape from the farm.

In August the folks took a trip with my aunt Betty and uncle Gerry to Fort Wayne, Indiana, leaving us kids behind. My sisters, the twins, and two-year-old Marty, I guess, went to the grandfolks. Left on the farm to do the chores, I talked Dan, who was no longer working, into staying with me. He and Hal were about to join the Army together, leaving me to my own destiny.

Aunt Betty and Uncle Gerry had quit farming some years before. After selling their hardware store in Loretto, they'd moved to Crystal, a northwest suburb of Minneapolis, where they became partners in a garage that serviced the rigs of long-distance truckers. The business also provided school buses for nearby Robbinsdale High School. The trip to Fort Wayne, by Greyhound, was to pick up two new buses at the factory there and drive them home to Crystal.

The two couples split up for the drive home, I heard afterwards, with Dad and Uncle Gerry in one bus, Ma and Aunt Betty in the other. They took turns driving and gabbed all the way. It was a fun little vacation for them.

Dan and I meant to have fun on the farm. He would help me with the chores, which weren't much (we'd finished second-crop alfalfa) except for the milking twice a day. We'd have plenty of time, we thought, for jaunts in the woods, for fishing and swimming, for watching TV or going to a show at night. As it turned out, after fooling around the first couple of days, we were called on by a neighbor who had been told by my father (who'd neglected to tell me) that I could give him a hand if he needed one. He had hay down, and looked pleased to find the two of us strong young boys available to help him.

He saved us, in effect, from ourselves. We both had .22 revolvers, obtained by mail order from Sears or Montgomery Ward, and had been practicing quick draws from our low-slung, tied-down holsters. Standing side by side, on a count of three, we'd both draw and fire; see who fired first. After several goes at this, I jammed my gun into its holster, caught the hammer, and had it fire down the inside of my jeans. Dan saw the spurt of dust beside my ankle; I felt a burning sensation. "You shot?" he said. I lowered my jeans and we both stared at the blue powder line down my leg, then laughed as if demented.

We played with dynamite. In the old milk house, now used as a tool shed, there was still a tin box of blasting caps, a roll of fuse, and a wooden box half-full of oily, sweating sticks of dynamite left over from the time my father and a hired man blasted the big shade tree out of the pasture below the barn so he could plow it up for crop land. My mother and us kids had stood out on the lawn to watch. We saw the men set the charge, then run for the barn. BOOM! and the huge old bur oak lifted out of the ground and fell over on its side. Dirt and rocks flew into the sky. Then crack! a big rock crashed through the barn roof and then a hail of stones and clods of earth pelted us on the lawn and clattered against the roof of the house.

I'd watched my father prepare a charge; knew how to attach a cap to a length of fuse, crimp it with the plier-like tool used for that purpose, then punch a hole at one end of a stick of dynamite and shove the cap and part of the fuse into it.

We took a loaded stick to the wooded ridge across the heifer pasture and I placed it under a fallen log, lit the fuse, and ran to the tree behind which Dan was already crouched. We waited for several minutes, it seemed. Then *BOOM*. The log lifted and broke in two in midair. "Good God!" Dan cried appreciatively.

Another thing that impressed Dan was how I could stand balanced behind the wheel of our new Allis Chalmers WD tractor and drive full throttle in road gear along the fence in our upper field to shoot at tin cans on the posts with my pistol. (The new Allis had replaced our old Oliver; we still had our old Farmall.) My second try, I turned too sharply at the end of the field and, because of its small, tricycle-like front wheel arrangement, nearly tipped the tractor over. That scared the recklessness right out of me. Sometime before this, one of the Miller boys, using *their* new tractor with the bucker on it as a battering ram, kept gleefully ramming into their straw pile until something broke. I never heard how old man Miller reacted to this, but I knew the supreme hell I'd have gotten from *my* old man if I'd damaged his new tractor.

Next morning the neighbor took us away from such foolishness and put us to work getting his hay in. I resented it at first; it was a hot, humid day, and hauling sixty-pound bales of alfalfa is laborious in any weather. Besides, the hay wasn't ours, and Dan and I might have been out on a cool lake, fishing. But then we started competing: he was a town kid, after all, anxious to show he could keep up with me, and I was a farm kid trying to show that he couldn't.

After his hay was in the barn, the neighbor brought out bottles of beer from the house and the three of us squatted under a shade tree on his lawn and drank them. That was his payment to us: treating us like men and to what men enjoyed after a day's work.

St. Thomas was far too expensive for me to attend another year, and eventually I'd be drafted if I stayed on the farm without the college student exemption. So my father was resigned once more to my leaving, and I was ready to join either the Air Force or the Navy (I *still* hadn't decided which), when we learned I could attend the University of Minnesota for only forty-three dollars a quarter. I might lose my religion in that pagan place, if not turn commie, but my practical father saw it as the way to keep me on the farm for at least another year.

He went with me to the university to talk with a counselor—I was touched by that, by his taking the time to venture with me into a world utterly foreign to him. The hugeness of the place was a little frightening at first after a year at small, comfortable, protective St. Thomas. But it also excited me with its thousands of students, its hundreds of attractive girls, its great rectangular mall between massive concrete buildings that formed the nucleus of the campus and was like the forum of an ancient city, a city of learning. To me the university was classical Athens or ancient Rome within the homely sprawl of Minneapolis.

In Advanced Writing I found another literary mentor in the stout, strict, kindly faced instructor I remember as Mrs. DelPlaine; her looks and pleasant voice—it had the sound of the prairie in it—put me in mind of Willa Cather. She was so conservative as to forbid the use of contractions in our writing. I submitted a piece to her on squirrel hunting, written in a self-conscious attempt at a style that was part Hemingway, part Hilaire Belloc, and part Mika Waltari's in his bestseller *The Egyptian* as translated from his original Finnish by Naomi Walford. It was mannered, in short, pseudo-biblical. I got a so-so mark from Mrs. DelPlaine, but she called me in to talk about my writing and to stress that I would have to get serious if I hoped to succeed at it and at college itself. She knew from my submissions to her that I was a farm boy. "Do you have time from your work to study at home?" she asked. Well, yes, I told her, at night after chores, but only a little on weekends when I worked full time for my father.

"Get your father to let you come into town on Saturday to study," she said. She suggested the Minneapolis public library as closer to my home than the university, and a good, quiet place to study.

Surprisingly, my father gave me the time off. After morning chores on Saturday, I'd wash and change my clothes, then drive in the family car to Minneapolis and the reading room of the city library on Tenth and Hennepin, where, among mostly old men—bums, I supposed, from off the street, enjoying the warmth as much as the reading matter—I studied to the quiet shuffling and crackling of newspapers around me. Usually I brought a bag lunch from home, which I could eat as I sat at one of the tables. For a break I'd browse the bookshelves or go upstairs to the little museum, which included an Egyptian mummy. Later I took to treating myself to a matinee movie. The theaters on Hennepin Avenue opened at 12:45, and you could see a show for eighty-five cents.

Toward the end of the fall quarter, I wrote a piece for Mrs. DelPlaine about the dilemma I faced on the farm; about the pressure I felt to stay on it and maybe inherit it someday when I didn't want it, about the guilt I felt for wanting to leave, about my sense of entrapment. She called me in to say I should submit it to *The Ivory Tower*, the weekly magazine edition of the *Minnesota Daily*, then swelled my head to bursting when she allowed that I might become a writer someday.

Dizzy with her encouragement, I went to the basement office of the *Daily*, where the editor said he might publish my piece if I could turn it into a story. I took it home, tacked on a scene in which the narrator, after failing to tell his father he wants to leave, resigns himself to staying, then goes out into the woods where since boyhood he's found temporary escape from the farm. I called it "The Woods," and the *Tower* accepted it.

Despite this forthcoming publication and her guarded prediction of my future, I got only a C from Mrs. DelPlaine for the first quarter of her course. But knowing her by now, and knowing *from* her a little of the vastness of what I must learn, her mark seemed fair enough.

January 1955. It was the start of the winter quarter at the U, and I had just passed my twentieth birthday. I was having lunch by myself (hot beef sandwich with mashed potatoes and gravy; hot fudge sundae for dessert, my favorite meal there) in the old Varsity Café on University Avenue at the entrance to Dinkytown, the commercial village on the edge of the Minneapolis campus. I was sitting in a corner booth among a noisy crowd of fellow students, and propped in front of me was Jack London's first collection of stories, *The Son of the Wolf*, which I'd checked out of the university library. I was reading the opening story, "The White Silence," about dog sledding through the pitiless Yukon wilderness. Outside the warm café, it was cold and snowing, yet another Minnesota winter. I sank into the story. I imagined writing it.

I had the rest of this school year—and then what? Escape from the farm finally, maybe—and the wide world in front of me. Some of London's success?

Anything seemed possible. I was that young.

Then, on a Friday night in mid-January, my uncle Billy, still serving as a recruiter in Minneapolis but about to retire after twenty-two years in the Navy, came out to the farm and, all at once, escape from the farm began staring me in the face.

The G.I. Bill, the Korean War G.I. Bill, which had been in place since the end of the war in 1953, would be canceled at the end of the month, my uncle told me. If I planned to finish college on the Bill, I'd better go in the service right away. He outlined the advantages of joining the Navy as if I'd stopped by his recruiting office to hear his pitch.

I spent a sleepless weekend thinking about it. Then on Monday, in a daze, as if powerless to do otherwise, I went down to the Navy recruiting office and enlisted for four years. That Friday, after passing my physical (despite occasional asthma, my lungs were clear; a "functional" heart murmur was detected), I withdrew from the university. The next day a story appeared, along with a picture of my uncle and me, in the Minneapolis *Star*. The headline read: RECRUITER'S SEA YARNS WIN NEPHEW TO DUTY WITH NAVY. I looked about fifteen. After that, things accelerated.

We had a family conference. The talk was of quitting farming. Mechanization was increasing to the extent that expensive bulk units were being forced on dairy farmers by state legislation and the big creameries. The units pumped milk directly from the cows into a bulk tank for transfer, a couple of times each week, into a tank truck for hauling to a creamery. The cost for such a unit, for our operation, would be some five thousand dollars, Dad figured. Twin City Milk Producers would install the unit on credit and deduct the cost, over the following months or years, from the folks' milk cheque. They'd owe their soul, in short, to the company store.

"We'll have to quit," Dad said.

Earlier my father had talked of increasing our milking herd from twenty-eight to fifty Holsteins. The folks might have continued farming, I suppose, if I hadn't been leaving, might have made it with the help of a hired man until my twin brothers were old enough to do a man's work. But since Nancy's death, they'd had the will for it knocked out of them. My mother especially had lost heart. My leaving, then, was only the final straw. Or so I told myself.

On the Monday before our auction, the neighbor woman who'd ridden with me to the university (I'd charged her the going rate of fifty cents a ride) stopped by on her way home. She was a pretty, tired-looking woman in her mid-thirties, a farm wife and mother whose children were all in school finally and who was working stubbornly toward a pharmacy degree at the university. She stopped to give me a copy of *The Ivory Tower* with my first published story in it.

Our auction was that Saturday. The next Monday I started boot camp.

The auction is a blur. I remember the portable diner, wheeled into our yard to provide food for those attending: hot dogs and hamburgers, french fries, pop or coffee. I see the line of cars and trucks pulled into our driveway, the knot of farmers, neighbors mostly, who showed up at eleven that morning to maybe bid on what Mervin Klatte had to sell. Vaguely I recall snatches of the auction itself, the procession of livestock, all our cows and heifers, brought out, one by one, to be bid on; recall the crowd's movement from one piece of our machinery to another, the auctioneer's barked delivery, the shouted bidding.

Holstein cow, fresh December 14
Holstein 3 year old heifer, due to freshen Feb. 25
Holstein cow, fresh Jan. 18, calf at side

1949 New Holland side rake, power takeoff type
New Burr feed mill
Corn binder, Electric fencer, Sunbeam clipper

Somebody bought my old Model A car for exactly what I'd paid for it: $35. Somebody else bought my mother's old upright piano, with my collected *Storys and Poems* (sic), in a three-hole binder, hidden for posterity behind the panel under the keyboard.

Mostly I remember my mixed feelings on that cold bright day at the end of January of 1955 as I watched the world of my formative years, the only world I knew then, breaking up before me. It was breaking up for all of us in my family, and we would feel its loss afterwards, as we felt the loss of little Nancy, whose accidental death is at the bittersweet core of our often fond and funny memories of the farm.

It was the week of the St. Paul Winter Carnival. I was in the King Boreas Company, an all-Minnesota company of Navy recruits. The Sunday after the auction, in connection with the carnival, the company was treated to dinner and patriotic speeches at the Capitol. I wasn't stirred by the rhetoric. I'd joined the Cold War Navy not to keep the world safe for democracy but just to see it. That night I lay sweating and sleepless on the couch in Uncle Billy and Aunt Rhonda's overheated livingroom in their house in Minneapolis. I was in space, floating between worlds.

Monday, January 31, 1955. Our company mustered in the St. Paul train station. In our civvies, with a band playing "Anchors Aweigh," we were set to marching, raggedly, down the platform and into the train. As we pulled out, I caught a last glimpse of my mother and father, my sisters and brothers, all smiling on the platform. There were tears in my mother's eyes.

On the train, I had no thoughts, only feelings. The car I was riding in, full of other boys like me, all of us leaving home for the first time, all heading into the unknown, was too raucous for thought. Thought would have been the enemy. We were being carried away, toward something strange and harsh and regimented, something too overwhelming to contemplate. And so we talked and laughed uproariously, shouting above our nervousness, as we rushed toward whatever awaited us.

❧

Within days of my leaving, the farm was sold. With money from the auction and from the sale of the buildings and the land (our home eighty, that is; the forty acres in Hamel—wisely, as it turned out—was retained), the folks paid up their bills, bought a house with an attached grocery store in Hamel, and had money left over to put in the bank. My mother ran the store, my father worked for Northland Creamery in Minneapolis (*If you can't fight'm, join'm, I always say*) and in his off hours, over the next five years and with the help of my twin brothers, turned our pasture and crop land below Hamel into a nine-hole public golf course. "Klatte's Folly," it was called by the town wags until they noticed all the cars turning off the highway into the course parking lot.

Later, after buying my uncle Lucien's adjoining twenty acres, my father added a "back" par-three. Eventually, with more land bought with the increasing profits from the course, he lengthened the "front" into a regulation nine. Finally, after his death in 1987, my three brothers bought more land still and created a regulation eighteen-hole golf course on pretty much the original holding of our great-grandparents.

The course has since been sold, so that last vestige of our farm, land that my great-great grandfather settled in the middle of the nineteenth century and that my father and I worked a hundred years later, is gone now—gone out of the family.

There are no farms around Hamel anymore, only strip malls and subdivisions, more and more of them each time I visit. And, yes, more golf courses, along with the malls and subdivisions, eating up what remains of the countryside. That country, the country of my heart, is all but gone now too, though it still exists in my heart and in my memory. So it is, I suppose, with those of us who have an abiding, even a helpless, sense of place.

In the heady days of the golf course's early success, I'd come home for the weekend from the university (to which, on the G.I. Bill, I'd returned after my release from the Navy) to find the folks piling and counting stacks of money on their bed, the week's "take." *Just look at that, Ross. It's mostly cash! Can you imagine us making that kind of dough on the farm?*

And he'd slip me a twenty.

And yet this poor boy during the boom of the Twenties, this survivor of the Depression Thirties, never trusted the way the money rolled in on the golf course. It was just too easy. It wasn't real. It was a bubble that could burst.

Hell, if times get tough again, we can always plow the course under and go back to farming. We'd eat, anyhow.

In the early evening, in the winter darkness, we arrived at the Great Lakes Naval Training Center and stepped off the train into the damp, Lake Michigan cold. We were met by our company commander, a tough little first-class petty officer, a submariner, we soon learned (after boot camp I would be assigned to the Pacific Submarine Force), who harangued us into a column of twos and marched us off the station platform and through the center's gate. "You'll be sor-ry!" sounded from an upper window of a gray barracks as we passed. "I want my maaama!" drifted down from another window. Sickening food smells wafted from another gray building, the chow hall, where we were headed. "Your *left*," the little sailor told us. "Your *left* right *left*."

Our column wavered as we lost step, then awkwardly regained it. We'd learn how to march, among other things, over our nine weeks of training. But I wasn't thinking of that, or of anything except the long stretch ahead of me of my Navy enlistment. I was already homesick; already my heart ached with nostalgia.

Four
> *years.*
Four
> *years.*

Those words beat to cadence in my swirling head as I marched toward the start of my other life, after the farm.

Acknowledgments

Thanks to the Canada Council for an Explorations grant that enabled me to start this project more than ten years ago. As for family members and friends who contributed their memories and encouraged me along the way, the list is far too long to include everybody, but these must be mentioned:

My parents, Mervin Ross and Ruth Marie Klatte; my sister, Joyce Bell; my sister, Marcia Baer, and my brother-in-law, James Baer, for reminding me of some farm practices; my brother, Mark, who witnessed the tragedy on our farm; my uncle, George Klatte, for his chronicle of my great-grandfather Williams's pioneering; my aunt, Estella Filkins, for her wry account of "the Major"; my cousin, Shirley Klatte, for her input on farming practices; my cousin, William Soderland, for his research into the Hamel family; my cousin, Rose Sleimers, for her genealogy of the Klatte family; my niece, Jamie Baer Petersen, for her interview with my mother; my aunt and uncle, Betty and Gerald Leuer, who once farmed with my parents; my maternal grandparents, Emily and Albert Hamel, and my paternal grandmother, Agnes Loomis, for their talk of old times; and not least, Sister Marjorie Rosenau, SSND, once our hired girl on the farm.

Thanks also to Ken McGoogan and Don Gayton, writer friends and exemplars; and to my old journalist friend, Allan Garske, for his critiques over the years. And to Hiro Boga, for her skillful editing. And to my loving wife of forty-plus years, April Lea, who supports my addiction to writing and suffers my crabbiness when the garbage needs taking out. And to our children, William and Alicia, who seem to enjoy their old man's reminiscences.

Ross Klatte was born in Minneapolis and grew up on his family's dairy farm just west of the city. After serving four years in the U.S. Navy as a journalist and obtaining a B.A. in journalism from the University of Minnesota, he worked as a reporter for the *Chicago Tribune*, as feature editor of the *National Bowlers Journal*, in Chicago, and as a copy editor for the *Detroit Free Press*. In 1971 he immigrated with his wife to Canada and homesteaded near Nelson, British Columbia. He worked there as assistant registrar of Notre Dame University, and later, in Student Services at Selkirk College. He and his wife still live outside Nelson and spend their winters in Mexico.

Ross Klatte's work has been published in literary journals, newspapers, and magazines. The opening chapter of *Leaving the Farm* is adapted from his original essay, which won the first prize in the CBC Literary Competition in 1990.

RECENT OOLICHAN TITLES

Father Tongue • Danielle Lagah
"Lagah's poems are beautiful, lucid stepping stones through the rivers of imagination that surround her south Asian heritage. A unique, inclusive journey through the world of emigration, difference and adaptation, written with exceptional clarity."—**Marilyn Bowering**

Notes for a Rescue Narrative • J. Mark Smith
"With this collection J. Mark Smith reveals himself as one of the most imaginative and accomplished young poets in Canada." —**Eric Ormsby**

The Blue Sky • Galsan Tschinag
"With this novel, a Mongolian shaman has stepped onto the stage of world literature."—*Der Spiegel*

Words • Mark Ellis
Illustrated by Ruth Campbell
"A stunning combination of word and image . . ."

Silent Inlet • Joanna Streetly
"A great book to curl up in a corner with . . . I loved it."
—**Phyllis Reeve**

To view our complete catalogue, visit us online at
www.oolichan.com